Principles of Chemical and
Biological Sensors

CHEMICAL ANALYSIS

A SERIES OF MONOGRAPHS ON
ANALYTICAL CHEMISTRY AND ITS APPLICATIONS

Editor
J. D. WINEFORDNER

VOLUME 150

A WILEY-INTERSCIENCE PUBLICATION

JOHN WILEY & SONS, INC.

New York / Chichester / Weinheim / Brisbane / Singapore / Toronto

Principles of Chemical and Biological Sensors

Edited by

DERMOT DIAMOND

School of Chemical Sciences, Dublin City University
Dublin, Ireland

A WILEY-INTERSCIENCE PUBLICATION

JOHN WILEY & SONS, INC.

New York / Chichester / Weinheim / Brisbane / Singapore / Toronto

Library of Congress Cataloging-in-Publication Data:

Principles of chemical and biological sensors / edited by Dermot
    Diamond.
        p.   cm. — (Chemical analysis ; v. 150)
    "A Wiley-Interscience publication."
    Includes bibliographical references and index.
    ISBN 0-471-54619-4 (alk. paper)
    1. Chemical detectors.   2. Biosensors.   I. Diamond, Dermot.
    II. Series.
    TP159.C46P75   1998
    681'.2—dc21   97-31765
                                                   CIP

Printed in the United States of America
10   9   8   7   6   5   4   3   2   1

For Tara, Danny, and Helen

CONTENTS

CHAPTER 7   **SENSOR SIGNAL PROCESSING**   **263**
                          Hugh McCabe

CONTRIBUTORS

John F. Cassidy, Department of Chemistry, Dublin Institute of Technology, Dublin, Ireland

Dermot Diamond, School of Chemical Sciences, Dublin City University, Dublin, Ireland

Andrew P. Doherty, School of Chemical Sciences, Dublin City University, Dublin, Ireland

Robert J. Forster, School of Chemical Sciences, Dublin City University, Dublin, Ireland

Gary Keating, School of Biological Sciences, Dublin City University, Dublin, Ireland

Anthony Killard, School of Biological Sciences, Dublin City University, Dublin, Ireland

Brian D. MacCraith, School of Physical Sciences, Dublin City University, Dublin, Ireland

Bernadette M. Manning, School of Biological Sciences, Dublin City University, Dublin, Ireland

Hugh McCabe, School of Electronic Engineering, Dublin City University, Dublin, Ireland

Teresa McCormack, School of Biological Sciences, Dublin City University, Dublin, Ireland

Richard O'Kennedy, School of Biological Sciences, Dublin City University, Dublin, Ireland

Francisco J. Sáez de Viteri, School of Chemical Sciences, Dublin City University, Dublin, Ireland

Johannes G. Vos, School of Chemical Sciences, Dublin City University, Dublin, Ireland

CHEMICAL ANALYSIS

A SERIES OF MONOGRAPHS ON
ANALYTICAL CHEMISTRY AND ITS APPLICATIONS

J. D. Winefordner, *Series Editor*

PREFACE

The development of chemical and biological sensors is an extremely dynamic and exciting area of scientfic research. It requires collaboration between scientists with widely differing disciplines, ranging from chemistry, biology, physics, and electronics. Teams with specialist expertise are required in analytical chemistry, synthetic chemistry, optics, electronics, signal processing, and instrumentaton, if devices capable of meeting the challenging specifications of end-users with real measurement problems are to be successfully developed. Given the tremendous breadth of activity in sensor-related research, it would be impossible to cover the entire area in a book of manageable size. We therefore have restricted the coverage to those areas which more or less coincide with current research activities at DCU (although these are also evolving as the emphasis in sensor research changes).

Over the past 10 years or so, a significant contribution to the development of chemical and biological sensors has been made by researchers based in Ireland. The idea for this book grew from a series of informal meetings to coordinate sensor-related research activities at Dublin City University (DCU) that resulted in the formation of the Sensors Research Centre (1989). Since that time, sensor research at Dublin City University has continued to flourish, and at the time of this writing involves some 40 full-time researchers working on a variety of externally funded projects. Welcome infrastructural support for our research has been provided through the Biomedical and Environmental Sensor Technology (BEST) initiative funded by the International Fund for Ireland and four universities active in sensor research; Dublin City University, The University of Ulster, Queen's University Belfast, and The University of Limerick.

I would like to acknowledge the tremendous effort of all the contributors to this text, both my colleagues at DCU, and their co-authors. I would also like to take this opportunity to thank all those in the research community who have contributed to the science of sensor development and our knowledge of this fascinating subject. Through joint projects, contributions to the literature or presentations/discussions at conferences, they have also contributed to this book.

Finally, my co-authors and I would like to acknowledge the help of our families during the preparation of this text. Without their support, this book would never have been completed.

DERMOT DIAMOND

Dublin City University
Dublin, Ireland

Principles of Chemical and Biological Sensors

CHAPTER

1

OVERVIEW

DERMOT DIAMOND

School of Chemical Sciences, Dublin City University, Dublin, Ireland

1.1. INTRODUCTION

Over the past 20 years or so, research and development (R&D) in the sensors area has expanded exponentially in terms of financial investment, number of papers published, and the number of active researchers worldwide. In part, this has occurred because of the links between developments in the semiconductor industry and the types of new devices under research, which are often based on semiconductor substrates. But in addition, it has also gained momentum from the enormous commercial potential for certain sensors, such as the hydrocarbon detectors used in the automobile industry, and sensors required for important clinical assays. Legislation has also opened up huge potential markets in environmental monitoring, for example, toxic gas and vapor monitoring in the workplace or contamination of natural waters by fertiliser runoff from fields. We will soon witness a quantum leap in terms of what may be termed the *performance–price index,* in that we shall see devices that perform much better than the previous generations of sensors and are less expensive to purchase. In other words, we are on the verge of a sensors revolution, similar to that of the microcomputer revolution of the late 1980s. The aim of this book is to introduce the main concepts and means of construction and use of the more popular types of chemical sensors available today, and to speculate on the future development of this fascinating area of scientific research.

1.2. SENSORS AND TRANSDUCERS

Before presenting any detailed discussions on sensors, let us initially try to define what we mean by a *chemical sensor.* In the literature, attempts have been made to discriminate between a chemical sensor and a physical transducer. The latter is popularly regarded as a device that provides an electronic signal through which changes in a particular property in its immediate environment can be monitored. Examples of transducers in common use include devices for temperature measurement or monitoring (thermocouples, thermistors, or platinum wires), piezo-electric pressure transducers, flowmeters, humidity sensors, light density sensors (photodiodes, charge-coupled devices, photomultipliers, photoresistors), and linear-rotational movement transducers. These devices often come close to meeting the general requirements of an ideal sensor in that they are sensitive, selective for the parameter of interest, cheap to purchase, physically robust, and easy to calibrate and use.

In addition, as most of these transducers are mass-produced, their behaviour is usually predictable, and the characteristics very reproducible within batches of the same device. They are available from many sources (Radionics, Farnell, Radio-Shack, etc.) and usually will have a data sheet available that provides the user with a complete character reference for the device, together with details of how to process the signal for a variety of common applications. Unfortunately, the situation with chemical sensors is not so advanced, and in many cases, devices are still handmade, with relatively large batch-to-batch irreproducibility in characteristics and attendant high production costs. This makes chemical sensors somewhat more difficult to use, and skilled staff are needed for accurate calibration and general troubleshooting.

1.3. CHEMICAL SENSORS

A chemical sensor can be distinguished from a physical transducer in that it provides the user with information about the chemical nature of its environment. While it may contain a physical transducer (e.g., a thermistor or piezoelectric crystal) or reference electrode (e.g., a Ag/AgCl wire) at its core, its overall character is usually determined by some type of chemically selective membrane, film or layer at the sensing tip. The composition and form of this layer is of crucial importance in determining the effectiveness of the sensor, as it controls the selectivity, sensitivity, lifetime, and response time of the device. Furthermore, the chemical recognition by this film of the analyte is often determined by a particular component of the film such as a ligand or particular functional group(s) that ultimately produces the signal, and that is therefore the actual sensor that defines the nature of the device. This is important if one is to distinguish bona fide chemical sensors from techniques such as open-cell FTIR (Fourier transform infrared), X-ray, fluorescence, and other spectroscopic methods that can be used to provide chemical information (e.g., remote mapping of elements in space, or waste emissions from industrial smoke stacks). The definition by Janata and Bezegh sums up the situation succinctly by emphasising both the role *and* the fundamental elements of these devices: "a chemical sensor is a device which furnishes the user with information about its environment; it consists of a physical transducer and a chemically selective layer" [1]. In addition, most users expect a sensor to be a small, probelike device capable of monitoring dynamic fluctuations in a target species in real time or providing a feedback signal for control purposes.

1.4. BIOSENSORS

Differentiation between biosensors and chemical sensors can be difficult. As with the chemical sensors–physical transducers division, the problem is one of whether to emphasise the role of the device or the nature of the recognition process involved in generating the selective signal. One definition of a biosensor is a sensing device that incorporates a biological entity (enzyme, antibody, bacteria tissue etc.) as a fundamental part of the sensing process. This definition obviously stresses the signal generation process. On the other hand, chemical sensors such as sodium or potassium ion-selective electrodes are often used to monitor these ions in biological fluids such as blood or urine. While these are not biosensors as defined above, they, and many other sensors, are used in bioanalytical situations. The term *bioprobe,* defined as a device which senses a species which is of biological origin, or a species which is an important component in a biological system, is sometimes used to distinguish these devices from bona fide biosensors. As this definition emphasises the application, it includes any sensor, transducer, or biosensor used for bioanalytical measurements. With many biosensors now being developed for non-biological applications, these terms provide a useful means of classifying sensing devices unambiguously.

1.5. CHARACTERISTICS OF AN IDEAL SENSOR

The first point that should be appreciated is that the "ideal" sensor does not exist—they all deviate from ideality to a greater or lesser degree. In addition, the acceptable characteristics are often a function of the application. Hence a sensor that performs well for monitoring a particular analyte in a given situation may be totally unsuitable for monitoring the same analyte in a different matrix. Factors that contribute to the suitability of a device for a particular assay may include physical parameters such as temperature, pressure, or even size, or chemical interferences arising from the sample matrix. Thus a glass electrode is an excellent device for monitoring pH in most everyday situations, but it is unsuitable for in vivo blood monitoring, for high-temperature pH measurements and is difficult to manufacture in microdimensions. Desirable characteristics for a sensor are listed in table 1.1, with some comments where relevant.

Table 1.1. Ideal Characteristics of a Sensor

Characteristic	Comments
Signal output should be proportional, or bear a simple mathematical relationship, to the amount of species present in the sample	This is becoming less important with the advent of on-device electronics and integration of complex signal processing options to produce so-called smart sensors
No hysteresis	The sensor signal should return to baseline after responding to the analyte
Fast response times	Slow response times arising from multiple sensing membranes or sluggish exchange kinetics can seriously limit the range of possible application and prevent use in real-time monitoring situations tection; can be improved by using sensor
Good signal-to-noise (S/N) characteristics	The S/N ratio determines the limit of detection; can be improved by using the sensor in a flow analysis situation rather than for steady-state measurements—S/N ratio can also be improved by filter or impedance conversion circuitry built into the device ("smart" sensor)
Selective	Without adequate selectivity, the user cannot relate the signal obtained to the target species concentration with any confidence
Sensitive	*Sensitivity* is defined as the change in signal per unit change in concentration (i.e., the slope of the calibration curve); this determines the ability of the device to discriminate accurately and precisely between small differences in analyte concentration

1.6.   TRENDS IN SENSOR RESEARCH

Some insight into recent trends in sensor research can be obtained from the numbers of papers being published per year. Analytical journals are useful indicators of systems that have been directly applied to solving real problems or that point the way to novel analytical techniques. A computer based search of the *Analytical Abstracts* database [1980–1996 (the database for 1996 was not complete at the time of writing)] revealed several interesting patterns. Before examining these trends in detail, the reader should be warned of potential pitfalls in this type of exercise, such as

- The use of a slight variant of a keyword can lead to large changes in the number of "hits." For example, the descriptors *optode* and *optrode* are both used by sensor researchers in an almost mutually exclusive manner. Consequently, important papers may be missed from a search if either is omitted (optrode is more common with 74 hits compared to 20 for optode). In addition, the use of wildcards such as the "*" character can allow other variations such as optrodes to be gathered in the search (using *optrode** as the keyword resulted in an increase from 69 to 95).

- Variations in the number of hits may reflect changes in the popularity of terms rather than any fundamental changes in the research being carried out. A good example of this is in the growth in popularity of the term *sensor,* which is becoming more fashionable in contrast to the term *electrode,* which, although still popular, is decreasing relative to the term *sensor,* even though the same research groups are often involved in the publications.

Figure 1.1 shows the number of hits for the terms *electrode** and *sensor** for the period 1980–1994. To normalize for changes in the total number of papers being abstracted each year, the data are presented in terms of percentage of total papers abstracted per year rather than in absolute numbers. Firstly, the proportion of papers involving electrodes grew from 3.5% to almost 10%, with the bulk of the increase occurring between 1982 and 1984. At the same time, there has been a steady increase in the number of abstracts found with the keyword *sensor** to about 6% by 1996. By 1993, the absolute number of sensor-related abstracts had risen to 866 compared to 1331 abstracts found using the keyword *electrode**. Furthermore, from 1993, the combined percentage of total papers in the database hit with the keywords *electrode** and *sensor** is around 16%. This represents a huge research effort, underlining the present importance of this research, and the future applications growing from it.

Figure 1.2 shows the number of hits for various subgroups of sensors, including ion-selective electrodes (ISEs), optical sensors,[1] amperometric, biosensor, acoustic, and solid-state sensors, as a percentage of the total number of sensor papers published each year. This clearly shows a striking decrease in the number of ISE papers, with a corresponding increase in the number of amperometric, biosensor, and optical sensor papers. Clearly there has been a steady decrease in ISE-related research from its complete dominance in the early 1980s to the current situation, in which papers on biosensors, amperometric sensors, and optical sensors are of increasing importance. It must also be remembered that ISE technology (at least for inorganic ions) is mature compared to other sensor types, and publications tend to be much more on the applications side rather than fundamental. How-

---

[1]Found with the following combination of keywords and logical operators [optode* or optrode* or (optical *and* sensor*)]

OVERVIEW

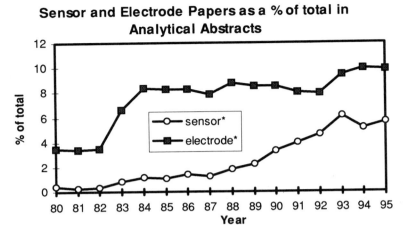

Figure 1.1. Papers abstracted using the keywords *sensor** and *electrode** expressed as a total of the number of papers in *Analytical Abstracts* from 1980 to 1995.

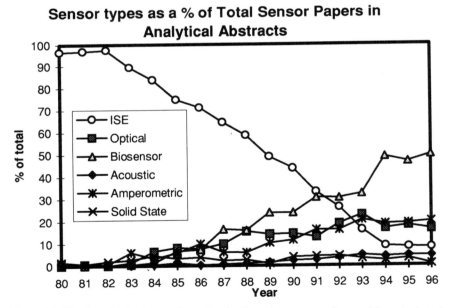

Figure 1.2. Trends in the absolute numbers of papers for each sensor type abstracted from *Analytical Abstracts* over the period 1980 to 1994.

ever, the same principles are being used to produce potentiometric sensors for more complex species such as drugs [2] and heparin [3], and these are seldom described as ion-selective electrodes in the literature. In addition, solid-state analogs of ISEs such as ISFETs (ion-selective field-effect transistors) or screen-printed potentiometric sensors, do not fall into the ISE category.

Trying to monitor optical sensor research activity in this manner is difficult as there is no single term (unlike the term *selective electrode* or *biosensor*) to access it as a coherent block. For example, optode returns 20 hits, optrode 74, whereas the combination [(*opto* or *optical**) and *sensor**] returns 779 hits (*Analytical Abstracts,* 1980–1995). Clearly there is a need for keyword definitions to be decided on for this important branch of sensor research. It should also be appreciated that much of the solid-state sensor research is being published in applied physics journals that are not abstracted into *Analytical Abstracts.*

1.7. THE SENSOR MARKETPLACE

A recent survey of the sensor marketplace [4] identified medical applications as being as being a major driving force for the development of emerging sensor technologies (fiber-optic sensors, smart sensors, silicon micromachined sensors, and thin-film devices).

The total market for fiber-optic sensors was $126 million in 1992, and with growth rates reaching 40% per year, this is predicted to reach approximately $1 billion by 1998, with the main impetus coming from blood chemistry analysers and intercranial pressure sensors.

"Smart sensors" are devices that incorporate certain electronic logic, control, or signal processing functions, and that therefore offer enhanced measurement capabilities, information quality, and functional performance. While these are found mainly in the physical transducer world, they are beginning to appear in chemical sensors [e.g., in ISEs containing impedance conversion circuitry and ADCs [analog-to-digital converters)]. Silicon micromachined devices are also mostly physical transducers and actuators, and the technology now exists to enable miniaturized sensors to be integrated with microflow systems incorporating valves, mixing chambers, and separation capabilities. The analytical possibilities arising from the integration of sample preparation, separation, and sensing in a single miniaturized unit has led to the concept of μTAS (micro–total analysis systems), which is fast becoming an area of analytical chemistry in its own right. Over the past few years, miniaturized planar or three-dimensional (3D) systems measuring a few square centimeters or less have been described that enable efficient separation and detection in much shorter times than can be achieved with conventional HPLC [5] or capillary electrophoresis (CE) [6] systems. The combination of small size and rapid analysis offered by μTAS instruments is blurring the traditional distinction between sensors and instruments as these new systems display features such as real-time measurements and small size, which are normally associated with sensors. The driving force for silicon micromachined sensor technology is coming from the huge demand for pressure sensors, flow sensors, and accelerometers from the biomedical and automobile markets. Sales of these devices are predicted to reach $2.2 billion by 1998.

1.8. CHALLENGES IN SENSOR RESEARCH

Progress in sensor development increasingly requires a multidisciplinary effort and access to more complex fabrication technologies. Traditional sensor manufacturing companies have tended to be small entities in terms of personnel and financial power, specializing in a few (usually handmade) products, and with low turnover and high profit margins. This situation is changing rapidly, and while the traditional markets for conventional probes will undoubtedly remain (e.g., pH glass electrodes), new products based on entirely different fabrication technologies will appear, developed by much larger companies with backgrounds in semiconductors, electronics, or printing rather than sensing.

Some idea of the diversity of research contributing to progress in sensor development can be gleaned from Figure 1.3. Particularly important areas include

1. Improving the recognition mechanism—this is of fundamental importance as it is the basis of the signal that will be obtained from the sensor. Researchers are synthesizing new molecular receptors with better selectivity, which respond to new target species, or that undergo a different transduction mechanism (or can provide information via several different transduction mechanisms). Materials under active research include the ion–molecule receptors (crown compounds, calixarenes, cyclodextrins, etc.) and enzymes–antibodies for biosensors. With the latter, major problems exist in retaining the activity of a biomaterial after immobilization in a sensing membrane or film, and in generating a reversible signal in the case of antibodies.

2. New materials are being investigated for use as a matrix in which to immobilize the receptor molecule. Important contributions are being made by polymer technologists who have developed new materials with attractive properties—easily handled as monomer (nonviscous), a wide variety of polymerization initiation mechanisms (e.g., photoinitiation, electropolymerization), rapid setting, and good stability in different environments. Some groups are investigating the possibility of generating selectivity in films through template effects by forming polymers around target molecules that are subsequently removed from the film, leaving their molecular imprint behind. Hence these films will have the correct conformation to

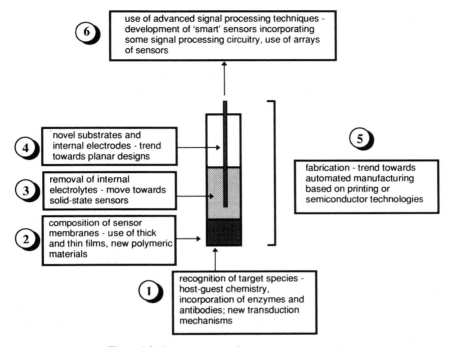

Figure 1.3. Important areas of current sensor research.

encapsulate these particular molecules from samples, and generate a response that can be monitored with appropriate instrumentation.

3: Sensors such as pH electrodes and ISEs contain an internal electrolyte in order to generate a stable internal reference potential. This restricts the design options for these sensors and the ability to automate fabrication. The generation of solid-state versions of these sensors is therefore an important challenge.

4. New sensor substrates (i.e., materials constituting the body of the sensor, or on which the device is built) are being investigated, arising mainly from a need for new planar fabrication designs.

5. There is great interest in adapting planar fabrication technologies for sensor manufacture with the goal of producing devices that are as similar as possible to each other, as this will hopefully lead to products with predictable characteristics (see below).

6. Improvements in signal processing technologies and instrumentation are making important contributions to the quality of sensor information. For example,

so-called smart sensors are being developed with a variety of signal processing options built in, ranging from impedance conversion to digitization and telemetry circuitry. In addition, the use of sensor arrays to probe samples, rather than single electrodes, can give important advantages such as multicomponent information, dynamic compensation for sample matrix effects, device malfunction–deterioration detection, and improved selectivity through the interpretation of response patterns obtained from the array.

1.9. PLANAR FABRICATION OF SENSORS

The driving force behind research into planar sensors has been the prospect of low-cost, large-scale manufacturing of identical devices. Techniques for depositing electrode materials such as silver, platinum, or gold on a micrometer or submicrometer scale such as chemical vapor deposition and low-pressure vapor deposition, are now used routinely in the microelectronics industry. With the use of appropriate masks, large numbers of microelectrodes can be patterned onto substrates. Precision liquid handling technologies including silkscreen or inkjet printing can be used to print electrodes onto substrates using "inks" containing the active components for the sensor. For example, inks containing silver and silver chloride are now available commercially. Films of controlled thickness can be formed on sensor substrates by spin coating. Langmuir–Blodgett troughs in principle allow monolayers of sensor materials with controlled orientation to be deposited. Control of deposition on the molecular scale is needed from a fundamental point of view, as it enables experiments to be designed that probe the recognition process at a molecular level.

Micrometer or nanometer scale sensor fabrication raises major problems in quality control—how to check the quality of a film that is not visible to the naked eye. Microscopy is necessary for this task, and progress in this area has been very rapid over the past 10 years or so, with the development of techniques such as scanning tunneling microscopy (STM), atomic-force microscopy (AFM), and confocal microscopy. While the instrumentation for these techniques is still relatively expensive, prices continue to fall. However, in combination with spectroscopic surface analysis techniques such as XPS (X-ray photoemission spectroscopy) and XRF (X-ray fluorescence), these techniques will become important routine methods for determining the quality of nanometer- or micrometer-scale sensor membranes and/or films. In addition, the instrumentation can be used with chemically sensitive nanometer-scale probes to provide new sensing techniques such as scanning electrochemical microscopy (SECM) or scanning potentiometry, which can provide information on the distribution of chemical species on suitable surfaces on a micron or sub-micron scale [7,8].

1.10.　INSTRUMENTATION AND SENSOR ARRAYS

Instrumentation has major influence on the commercialization of sensors. A good illustration of this was the rapid expansion in the pH monitoring market following the development of the field-effect transistor (FET) and its incorporation into operational amplifier integrated circuits in the early 1960s, which made it possible, for the first time, to make small, low-power, portable instruments that could measure high-impedance voltages very accurately. An identical situation is currently developing in the field of integrated optics, which is leading to the development of small (pocket or even chip-sized) UV-vis (visible) and near-IR spectrometers. These, coupled with novel optically responsive films deposited on waveguides and the availability of cheap, portable PCs, present a powerful driving force for new optical sensor-based applications for field measurements. In electrochemical measurements, the digitization of sensor signals has become an important factor. Almost all current/voltage measurement instrumentation these days is digital in nature, with built-in analog-to-digital converters (ADCs) and computer interface ports (usually serial RS-232).

PC-compatible interface cards capable of monitoring up to 16 analog voltages have been available since the early 1980s but have not been used much, even in research. However, recently interest has been growing in sensor arrays (Figure 1.4), as there is potentially much more information in the pattern of signals obtained from an array compared to the traditional single-sensor approach. Perhaps most importantly, a well-characterized sensor array can respond dynamically to changes in a sample matrix, and compensate for possible errors arising from lack

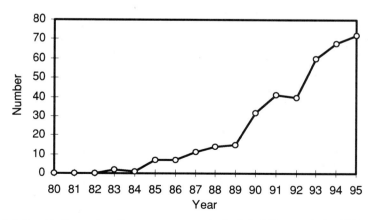

Figure 1.4. Trend in the number of sensor array papers in *Analytical Abstracts* during 1980–1994.

Figure 1.5. PCMCIA input/output (I/O) board for portable PC. The multifunctional DAQcard 1200 from National Instruments slots into one PCMCIA slot converting the portable PC into a powerful data acquisition and control system.

of selectivity. For example, commercial ammonium electrodes are not selective against potassium, and a user has no means of knowing whether a signal obtained from an ammonium electrode is dominated by ammonium, or potassium, or some mixture of both ions. However, if an array containing ammonium, potassium, and other sensors is used, a computer-based instrument can be trained to distinguish between the above situations to a certain degree [9], and hence increase user confidence in the measurements (see Section 2.10 and Figure 2.16).

Multichannel data acquisition cards based on PCMCIA (Personal Computer Memory Card International Association) technology (Figure 1.5) for portable PCs are now becoming available that convert these lightweight computers into powerful instruments with multichannel voltage measurement capabilities (see catalogs from National Instruments, Austin, Texas for details). Coupled with modem and telemetry products that are being generated and improved at an increasingly rapid rate, this new, lightweight, portable instrumentation will play an important role in expanding the range of applications for sensors, particularly in remote environmental or clinical monitoring situations. While these multichannel voltage monitoring products are readily adaptable for use with potentiometric sensors, commercial multichannel potentiiostats are only now beginning to appear. Researchers interested in developing voltammetric sensor arrays have, up to now, had to design and fabricate their own instruments, and this has limited progress with these systems. The ability to obtain real-time voltammograms in flowing streams via an array of individually controlled working electrodes opens additional dimensions of information (Figure 1.6) compared to single- or dual-channel amperometric measurements, which are the current norm in chromatography. The improved reliability of these measurements, along with the availability of new commercial products in the data acquisition and communications fields, is likely to further stimulate research in sensor arrays, as further applications are developed. Arrays can be of the same types of sensors (e.g., all potentiometric), with or without redundancy (i.e., more than one sensor in the array of the same type targeted at the same analyte) or of mixed sensor types. Figure 1.7 illustrates some advantages of sensor redundancy. It shows the display obtained from a four-channel biosensor array comprising two glucose and two lactate biosensors to spiked whole-blood samples. The array is monitoring the dialysate of the blood samples obtained via a microdialysis loop. In Figure 1.7a, all four channels are functioning, and the same response is obtained from the paired biosensors. However, in Figure 1.7b, one of the lactate sensors is clearly not giving a signal. As the other three sensors are functioning, it is not likely that the fault lies with the lactate biosensor, but rather with the microflow system. The fault was quickly traced to a leak in the flow system that prevented the dialysate from reaching this lactate sensor. This fail-safe feature, which is available through the duplication of the lactate and glucose sensors, was recommended by clinicians associated with the development of this system. The miniaturized instrument and microdialysis sampling unit has recently been successfully applied to on-line in vivo lactate and glucose monitoring in anaesthetized dogs [10]. Results obtained for glucose are illustrated in Figure 1.8. In this case, the use of a "blank" sensor (i.e., one that does not contain glucose oxidase) in the array enables noise artifacts to be easily distinguished from the analytical component of the signal. These artifacts are associated with manual manipulation of the sampling shunt during the measurements and can be easily substracted from the glucose sensor output to give a more accurate display of the true signal.

It is also possible to obtain several channels of complementary information from a single sensor. For example, electrochemical sensor films can be deposited on semitransparent electrode substrates to provide simultaneous electrochemical, spectroscopic (IR/UV–vis), and possibility acoustic [using a quartz crystal microbalance (QCM) instrument] information on a particular membrane process. However, not all sensor applications in the future will require these relatively complex approaches. In fact, a large proportion of sensor applications will continue to rely on very cheap, single-shot, disposable sensors that will be in competition with visual test strips rather than analytical instruments. What is clear, is that research, development, and the commercialisation of sensor technology will accelerate over

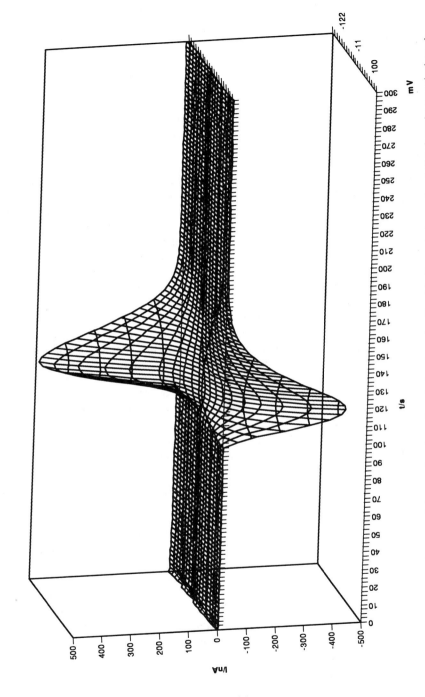

Figure 1.6. Response of an array of glassy carbon electrodes to hydroquinone in a flow-injection analysis experiment. The electrodes are poised at voltages increasing from −200 to +100 mV vs. SCE in steps of ca. 16 mV, enabling 3D voltammograms (*I* vs. *V* vs. *t*) to be monitored in real time.

14

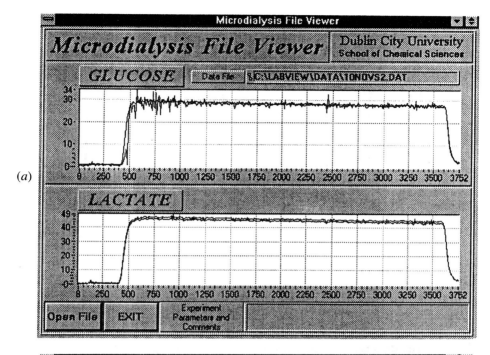

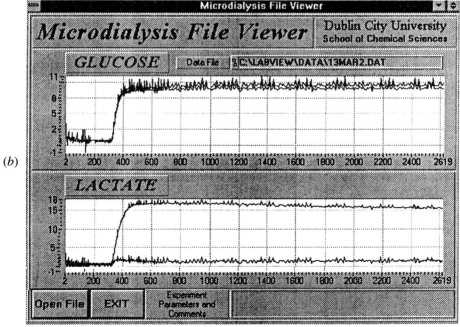

Figure 1.7. Microdialysis of serum sample through aminocellulose shunt (100-mm length): (*a*) all sensors working; (*b*) lactate 1 sensor not responding.

15

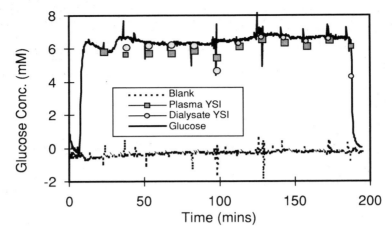

Figure 1.8. Analysis of glucose in blood using a biosensor array and microflow system to monitor the dialysate in real time. Good agreement between sensor-based system and the reference method is evident (Yellow Springs Instruments glucose analyzer). Close agreement between the YSI results in dialysate and plasma is convincing evidence of a high recovery through the dialysis membrane (dialysis glucose is an accurate reflection of the blood glucose). Despite the slow flow rate employed (10 μL/min), a very fast response is evident, with the steady state reached within a few minutes. This is a feature of the very small dead volume of the microflow system [10].

the coming years, and we can look forward to exciting new developments in this very dynamic area of scientific research.

REFERENCES

1. Janata, J., and Bezegh, A., *Anal. Chem.* **60,** 62R (1988).

2. Zhang, Z. R., and Yu, R. Q., *Anal. Chim. Acta.* **285,** 81 (1994).

3. Fu, B., Bakker, E., Yun, J. H., Yang, V. C., and Meyerhoff, M. E., *Anal. Chem.* **66,** 2250 (1994).

4. Data in this section are quoted from *Sensor Market Sourcebook,* 1995 Edition, product code 2811-32, Frost and Sullivan, March 1995, *Trends and Forecasts for the World Sensor Marketplace,* New York/London, 1995 edition.

5. Effenhauser, C. S., Manz, A., and Widmer, H. M., *Anal. Chem.* **67**(13), 2284–2287 (1995).

6. Ocvirk, G., Verpoorte, E., Manz, A., Grasserbauer, M., and Widmer, H. M., *Anal. Meth. Instrum.* **2**(2), 74–82 (1995).

7. MacPherson, J. V., Beeston, M. A., Unwin, P. R., Hughes, N. P., and Littlewood, D., *J. Chem. Soc., Faraday Trans.* **91,** 1407 (1995).

8. Arca, M., Bard, A. J., Horrocks, B. R., Richards, T. C., and Treichel, D. A., *Analyst* **119,** 719 (1994).

9. Saez de Viteri, F. J., and Diamond, D., *Analyst* **119,** 749 (1994).

10. Freaney, R., McShane, A., Keavney, T. V., McKenna, M., Rabenstein, K., Scheller, F. W., Pfeiffer, D., Urban, G., Moser, I., Jobst, G., Manz, A., Verpoorte, E., Widmer, M. W., Diamond, D., Demsey, E., Saez de Viteri, F. J., and Smyth, M., "Novel Instrumentation for Real-Time Monitoring Using Miniaturised Flow Cells with Integrated Biosensors," *Ann. Clin. Biochem.* **34,** 291–302 (1997).

SUGGESTED FURTHER READING

Books

1. Turner, A. P. F., Karube, I., and Wilson, G. S., eds., *Biosensors, Fundamentals and Principles,* Oxford University Press, Oxford, 1987: although a little aged now, still a superb compilation of papers covering all types of biosensors and their applications.

2. Seiyama, T., and Yamazoe, N., eds., *Chemical Sensor Technology,* Elsevier, Amsterdam, 1988–: a series of volumes focusing mainly on solid-state sensors and advances in fabrication technology.

3. Janata, J., *Principles of Chemical Sensors,* Plenum Press, New York, 1989: an excellent summary of the basic chemical principles underlying the response of the main types of sensors.

4. Wolfbeis, O. S., ed., *Fiber Optic Chemical Sensors and Biosensors,* CRC Press, Boca Raton, FL, 1991: multiauthor text covering all types of chemical sensors and biosensors based on optical transduction. Individual contributions vary from applications areas (biomedical, environmental, nuclear plantss), to specific analytes (pH, oxygen) and fabrication technologies.

5. Krohn, D. A., *Fiber Optic Sensors: Fundamentals and Applications,* 2nd ed., Instrument Society of America, Research Triangle Park, NC, 1992: covers the principles of optical sensors with main emphasis on physical transducers for temperature, level sensing, rotation, pressure, flow, and other parameters.

6. Madou, M. J., and Morrison, S. R., *Chemical Sensing with Solid State Devices,* Academic Press, Boston, 1989: an excellent text covering the fundamental principles of most sensor types, and focusing particularly on the problems involved in fabricating solid-state sensors.

Journals

Sensors and Actuators: dedicated to solid-state sensors and divided into two subseries. *A*—physical transducers and *B*—chemical sensors.

Biosensors and Bioelectronics: good source of papers dealing with developments in biosensors and associated instrumentation.

Analytical Chemistry: the biannual review series provide a very useful guide to the analytical literature in sensors over the previous 2 years; next review will be in 1998; regular papers on sensors in the general research section.

Analytical journals such as *Analytica Chimica Acta* and *The Analyst* and *Talanta* have sensor subsections and occasional special issues completely dedicated to sensor research papers.

CHAPTER

2

ION-SELECTIVE ELECTRODES AND OPTODES

DERMOT DIAMOND

School of Chemical Sciences, Dublin City University, Dublin, Ireland

FRANCISCO J. SÁEZ DE VITERI

Biomedical and Environmental Sensor Technology Centre, Dublin City University, Dublin, Ireland

2.1. INTRODUCTION

As a general definition, the term *electrochemistry* refers to the branch of chemistry that studies chemical reactions and processes in which electric charges are involved, either applied to produce chemical reactions or generated as a result of them. These processes, which can be very different in nature, have been used in the analytical field for many years and have led to the development of very distinctive techniques such as potentiometry, voltammetry, amperometry, coulometry, and more recently scanning electrochemical microscopy (SECM) [1,2].

Potentiometry, which is the focus of this chapter, deals with the electromotive force (emf) generated in an galvanic cell where a spontaneous chemical reaction is taking place. In the last 20 years or so this technique has widened its base of application as new techniques, devices, and instrumentation were being developed. Analytical potentiometry deals with the use of the emf response of a potentiometric cell to determine concentrations of analytes in samples. The glass-membrane electrode was the first ISE to be discovered and characterised, and its exquisite selectivity for H^+ ions and chemical robustness made it widely used for pH measurements. The technique rapidly became very popular, and different glass compositions were used to induce selectivity to other ions. Solid-state, gas-sensing electrodes or the liquid and polymer membrane electrodes are sensors that may be based on, or incorporate, an ion-selective electrode. Enzymes and proteins are also used as selective materials in ISEs for biochemical species as part of special electrode designs [3].

Apart from the continuous development and improvement of the recognition chemistry of these sensors, techniques to enhance the sensor performance, and robust response theories to explain their behavior have been elaborated, thus increasing the range of applications and giving a better understanding of the principles governing the signal generation process.

Numerous applications appear continuously in the literature describing new selective materials, new techniques, and new fields of application for potentiometric electrochemistry involving ISEs, in both the chemical [4–7] and biochemical [8] fields, and has been the dominant area of sensor research of the 1970s and 1980s, with over 200 papers being published annually.

2.2. POTENTIOMETRY

It is known that when a metal M is immersed in a solution containing its own ions M^{z+} an electric potential is developed on the metal surface as the following process occurs:

$$M^{z+}_{(aq)} + ze^- \rightleftharpoons M_{(s)} \tag{2.1}$$

This generates an electrode potential (E) that depends on the activity of the metal ions in solution and the number of electrons involved in the process, and that can be expressed as

$$E_{elec} = E^0_{elec} + \frac{RT}{zF} \ln a_{M^{z+}} \tag{2.2}$$

where E_{elec} is the electrode potential in volts, R is the gas constant, T is the absolute temperature, F is the Faraday constant, $a_{M^{z+}}$ is the activity of the metal ions in solution, z is the number of electrons involved in the process, and E^0_{elec} is the standard electrode potential, which is the value of the potential the electrode presents when $a_{M^{z+}} = 1$. At 25°C, and correcting from natural to decimal logarithm, Eq. (2.2) takes the form

$$E_{elec} = E^0_{elec} + \frac{59.16}{z} \log a_{M^{z+}} \tag{2.3}$$

also known as the *Nernst equation,* which states that, for a 10-fold or decade change in the activity of a monovalent cation ($z = 1$), a theoretical potential change of +59.16 mV should be expected. This potential change can be measured only against a reference electrode which presents a stable potential, provided the ionic strength and the sample matrix remain constant, and completes the electrochemical cell:

$$E_{cell} = E_{elec} - E_{ref} \tag{2.4}$$

Experimentally, changes in the cell potential are assigned solely to changes in the ISE potential as the reference electrode potential is regarded as invariant:

$$\Delta E_{cell} = E_{cell(2)} - E_{cell(1)} = E_{elec(2)} - E_{elec(1)} \qquad (2.5)$$

assuming E_{ref} = constant. In turn, provided the ISE response is Nernstian, these changes in cell potential can be related to changes in the analyte ion activity on transferring from solution 1 (a_1) to solution 2 (a_2) via Eq. (2.3):

$$\Delta E_{cell} = \Delta E_{elec} = S \log \left(\frac{a_2}{a_1} \right) \qquad (2.6)$$

where S is the electrode slope ($59.16/z$ mV per decade). The hydrogen electrode is the standard reference electrode, but more practical secondary reference electrodes such as calomel or silver/silver chloride are usually employed [9]. Membranes that selectively exchange ions (e.g., due to the presence of immobilized ligands) generate potentials in a similar manner, and the change in membrane potential on transferring between two solutions of the analyte ion is also described by Eq. (2.6) (see Sections 2.3 and 2.5).

2.3. THE NIKOLSKII–EISENMAN EQUATION

Ion-selective electrodes (ISEs) are potentiometric sensors characterized by the fact that the electrochemical response is usually dominated by one ionic species present in the solution, known as the *primary ion* (analyte ion, target ion). However, other ions, known as *interfering ions,* may also contribute to the membrane potential.

Figure 2.1 shows schematically a typical membrane-based ion-selective electrode/reference electrode galvanic cell.[1] Three different parts must be distinguished in the membrane: the two boundary or outer surfaces and the bulk of the membrane. When the membrane is placed between two solutions containing an ionic species k at two different levels of activities (a_k' and a_k''), a potential E_M is created across it. Assuming that (1) the only driving forces involved in the process are concentration and electrical forces—no temperature or pressure gradients exist across the membrane; (2) the membrane is in thermodynamic equilibrium, internally and with the two solutions at both sides; (3) both solutions are prepared in the same solvent (almost invariably water); and (4) the standard redox potentials of all the chemicals remain constant with time and space within the membrane, the

---

[1]In the term E_B the double prime (E_B'') refers to a solution of constant (fixed) ion activity that is equivalent to the electrode's internal electrolyte, while the single prime (E_B') refers to a solution of variable activity equivalent to the sample solution.

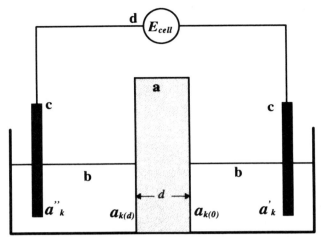

Figure 2.1. Schematic diagram of a selective membrane cell: (*a*) selective membrane; (*b*) aqueous solutions containing ionic species; (*c*) reference electrodes; (*d*) high-impedance voltammeter. See text for explanation of ion activities.

overall membrane potential (E_M) can be described, in an ideal membrane, as the sum of the boundary potentials E'_B and E''_B (outer surfaces), and the diffusion potential E_D (bulk of membrane) [10] as in

$$E_M = (E'_B - E''_B) + E_D \qquad (2.7)$$

where

$$E'_B = \frac{RT}{z_k F} \ln \frac{k_k a'_k}{a_{k(0)}} \qquad (2.8)$$

$$E''_B = \frac{RT}{z_k F} \ln \frac{k_k a''_k}{a_{k(d)}} \qquad (2.9)$$

and

$$E_D = \frac{z_k u_k \,(a_{k(d)} - a_{k(0)})\, RT}{z_k^2 u_k \,(a_{k(d)} - a_{k(0)})\, RT} \ln \frac{z_k^2 u_k a_{k(0)}}{z_k^2 u_k a_{k(d)}} \qquad (2.10)$$

Overall

$$E_M = \left(\frac{RT}{zF} \ln \frac{a_k - a_{k(d)}}{a_k - a_{k(0)}} \right) + \left(\frac{z_k}{z_k^2 u} \frac{RT}{z_k F} \ln \frac{a_{k(0)}}{a_{k(d)}} \right) \qquad (2.11)$$

where $a_{k(d)}$ and $a_{k(0)}$ are the activities of the primary ion k in the membrane phase at the internal and external boundaries, respectively; a_k'' and a_k' are the activities of the ion k in the internal and external solutions, respectively; and u_k is the mobility of k in the bulk of the membrane. For the ideal case where $a_{k(d)} = a_{k(0)}$, and assuming identical mobilities for all the similar species permeating into the membrane (i.e. $E_D \Rightarrow 0$ or it is constant), the expression for the membrane potential is reduced to

$$E_M = \frac{RT}{zF} \ln \frac{a_k'}{a_k''} \qquad (2.12)$$

In a disposition where the k^{z+}-selective membrane is the sensing membrane of an electrode j (Figure 2.2), the activity of the ionic species at one side of the mem-

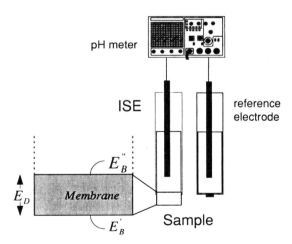

Figure 2.2. Permselective membrane electrode disposition. The selective membrane presents three different potentials, between the sample and the external surface or external boundary potential (E_B'), between both membrane surfaces or diffusion potential (E_D), and an internal boundary (E_B'') potential between the membrane and the internal reference electrolyte.

brane remains constant (electrode's internal solution of an ion activity a_k''), and Eq. (2.12) reverts to the Nernst equation [Eq. (2.2)]:

$$E_j = \frac{RT}{zF} \ln \frac{1}{a_k''} + \frac{RT}{zF} \ln a_k' = E_{elec}^0 + \frac{RT}{zF} \ln a_k' \qquad (2.13)$$

Note that E_{elec}^0 includes the internal membrane boundary potential, the membrane diffusion potential, and the internal reference electrode potential. If the sample solution contains a permeating interfering ion l of the same charge as the primary ion k, the membrane potential [Eq. (2.12)] can be expressed as

$$E_{j_k} = \frac{RT}{zF} \ln \left(\frac{u_k k_k a_k' + u_l k_l a_l'}{u_k k_k a_k''} \right) = E_{j_k}^0 + \frac{RT}{zF} \ln \left(a_k' + \frac{u_l k_l}{u_k k_k} a_l' \right) \qquad (2.14)$$

where k_k and k_l are the distribution coefficients of the ionic species k and l between the aqueous and membrane phases and a_l' is the activity of the interfering ion l in the sample solution. The ratio $u_l k_l / u_k k_k$ effectively defines the contribution of the interfering ion l to the overall electrode response, and it is usually represented as the potentiometric selectivity coefficient $K_{j\,kl}$. A more generalised expression, known as the *Nikolskii–Eisenman equation*, includes contributions from a number of interfering ions l of different charges z_l in the same solution i:

$$E_{ij} = E_j^0 + S_j \log \left(a_{ik} + \sum K_{jkl} a_{il}^{z_k/z_l} \right) \qquad (2.15)^*$$

where E_{ij} is the potential measured in the electrode j for a given solution i, E_j^0 is the standard electrode (half-cell) potential, S_j is the electrode slope, a_{ik} is the activity of the primary ion k and a_{il} is the activity of any interfering ion l in the same sample i, z_k and z_l are the charges on the primary and interfering ions, and $K_{j\,kl}$ is the selectivity coefficient of the electrode j against the interfering ion l, and it is defined as

$$K_{jkl} = \frac{u_l k_l}{u_k k_k} = \frac{u_l}{u_k} (_e K_{jkl}) \qquad (2.16)$$

where u_l and u_k are the ion mobilities in the membrane phase and $_e K_{j\,kl}$ is the equilibrium constant [9] of the reaction

$$l_{(aq)}^{z_l+} + k_{(mem)}^{z_k+} \rightleftharpoons l_{(mem)}^{z_l+} + k_{(aq)}^{z_k+} \qquad (2.17)$$

---

*The introduction of the electrode identifier j is required for a coherent approach including arrays of ISEs where the identity of the electrode is also a variable.

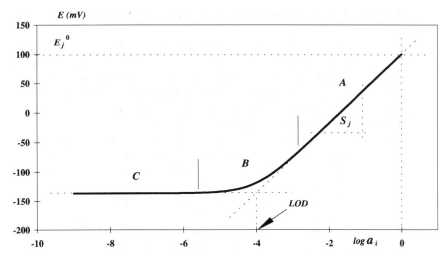

Figure 2.3. Example of the theoretical response of an ideal ISE. The curve represents the Nikol-skii–Eisenman response for an ISE with the following parameters: $E_j^0 = 100$ mV; $S_j = 59.16$ mV per decade. Overall contribution of the summation term equals 10^{-6} mol/dm^3. Segments A, B, and C correspond to case 1, case 2, and case 3, respectively, described in Section 2.3.

The summation factor $\Sigma_l K_{jkl} a_{il}^{z_k/z_l}$ in Eq. (2.15) is the error arising from the net contribution of all interferents. Figure 2.3 shows the theoretical response of the Nikolskii–Eisenman semi-logarithmic equation for a primary ion activity range of 10^{-9} to 10^0 mol/dm^3, with an arbitrary total contribution from the summation factor of 10^{-4} mol/dm^3 ($E_j^0 = 100$ mV, $S_j =$ Nernstian).

From Eq. (2.15), the most useful ISEs are clearly those that are very selective ($K_{jkl} \Rightarrow 0$) against a wide range of common interfering ions as for these electrodes $\Sigma_l K_{jkl} a_{il}^{z_k/z_l} \Rightarrow 0$, and hence changes in the measured potential (E_{ij}) can be related with confidence to variations in the primary ion activity alone. Three situations can be differentiated in terms of the relative values inside the logarithm in Eq. (2.15):

- **Case 1:** $a_{ik} \gg \Sigma_l K_{jkl} a_{il}^{z_k/z_l}$. This is the ideal case in which the contribution of the sample interferents l is negligible in comparison to the primary ion k. Graphically, it can be related to the ascendant part of the Nikolskii–Eisenman response curve in Figure 2.3, which rises with constant slope. In this case, analysis of a sample can be directly performed with a calibrated ISE, relating the electrode potential to the primary ion activity as in the following equation:

$$a_{ik} = 10^{E_{ij} - E_j^0 / S_j} \tag{2.18}$$

Calibration is required to predetermine E_j^0 and S_j (two-point calibration minimum).

- **Case 2:** $a_{ik} \approx \Sigma_l K_{jkl} a_{il}^{z_k/z_l}$. In situations where the net contributions of the summation factor are of the same order as the contribution of the primary ion, the approach used in Case 1 cannot be used for analysis unless the value of the summation term for interferent contribution is known and remains constant both in standards and unknown samples. In this case, the standards must contain the same constant background of interferent ions. In Figure 2.3, this case is represented by the curved portion of the theoretical electrode response. The calibration curve obtained with this series of standards contains the information and correction for the effect of this particular interferent background, and accurate prediction of unknowns with this calibration curve can be guaranteed only for samples with the same background composition. In practice, analysis in this region of the Nikolskii–Eisenman response profile is not used because the nonlinear response produces a variable slope even if the contribution of the summation factor remains constant. Besides, in this area, the value of the electrode sensitivity (i.e., the slope) is much reduced, decreasing the efficiency of the analysis.

- **Case 3:** $a_{ik} \ll \Sigma_l K_{j_kkl} a_{il}^{z_k/z_l}$. In this case accurate determination of the primary ion is not possible, as its response is masked by the contribution from interferents. Separation and preconcentration techniques such as ion chromatography [11] or gas diffusion [12] processes must be used prior to the determination of the target species. The part of the Nikolskii–Eisenman response with zero slope in Figure 2.3 represents this third case.

2.4. TYPES OF ISEs

In this section, an overview of the most common electrode types is presented. All of them are considered as ISEs, as they are potentiometric devices selective for ionic species. Other more complex designs exist that combine some characteristics from different sensor types, but cannot be included with the scope of this review. Alternative reviews [4–8] contain large number of references covering these novel ISE dispositions for particular applications.

2.4.1. Glass Electrodes

Glass pH electrodes have been widely used since the early 1930s to quantify the presence of H^+ in solution. These electrodes operate on the fact that when a glass membrane is immersed in a solution containing hydrogen ions an ion-exchange mechanism with the fixed SiO^- groups in the glass membrane boundary region is initiated [13]. The glass material is made of a solid silicate matrix within which alkaline metal ions are mobile. When this glass membrane is brought in contact with an aqueous solution, its surface becomes hydrated to a depth of about 100 nm and the alkali metal cations (principally Na^+) from the glass matrix can be exchanged for other ions in the solution, preferably H^+, creating a potential across the membrane that is a linear function of the pH of the solution.

$$Na^+_{(glass)} + H^+_{(aq)} \rightleftharpoons H^+_{(glass)} + Na^+_{(aq)} \tag{2.19}$$

Doping the glass membrane with different proportions of aluminum oxide and other metal oxides can produce glass membrane electrodes selective for other metallic ions such as lithium, sodium, potassium, silver, or ammonium [14].

2.4.2. Membranes Based on Crystalline Materials

The glass membrane from the glass electrode can be replaced with other solid materials to produce electrodes with selectivity for other ions. The two main types are crystalline and pressed-pellet electrodes.

In the first type, single crystals can be used as selective materials, as is the case of the fluoride ion-selective electrode which is based on a lanthanum fluoride (LaF_3) crystal doped with europium. This crystal serves as the selective membrane that conducts F^- ions at room temperature. When this membrane is placed in a electrode disposition, separating the internal reference solution (usually NaF mixed with NaCl) and the internal reference electrode from the sample solution, the LaF_3 crystal responds almost specifically to the activity of fluoride ions in the sample between pH = 3 and pH = 8. Outside this range H^+ and OH^- interfere by formation of HF, and by chemical modification of the structure of the LaF_3 crystal, respectively.

Other crystalline materials can also serve as selective membranes when pressed into disks or dispersed in suitable polymers, such as silver sulfide (Ag_2S) disks, which respond to silver ions due to the mobility of the silver ions within this ionic conductor. These electrodes are based on the low solubility of Ag_2S ($K_{sp} = 10^{-51}$) as, in an electrode disposition, the membrane creates an electrochemical potential between the internal reference and the sample solution, which depends on the activity of Ag^+ in the sample solution. Same principle can be adapted to respond to other ions such as like Br^-, I^-, Pb^{2+}, and Hg^{2+}.

Silver halide cast pellets can also be used as selective material for the corresponding halide ions but, usually, the single crystal version of these electrodes perform better. These constitute quite an important fraction of the commercially available ion-selective electrodes.

2.4.3. Gas-Sensing Electrodes

These electrodes often consist of a glass pH electrode covered with a hydrophobic gas-permeable membrane (GPM) and buffer solution filling the space between

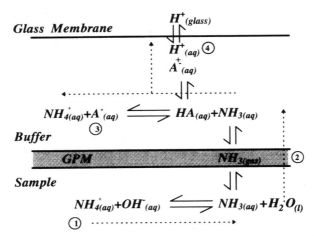

Figure 2.4. Chemical equilibria involved in an ammonia-selective gas-sensing probe. See text for details on the equilibria and gas diffusion mechanism. (GPM = gas-permeable membrane.)

the electrode and the membrane. When the membrane is immersed in a solution containing acidic/basic gaseous species (e.g., NH_3, CO_2, SO_2, NO_2), they diffuse through the membrane, dissolving in the buffer solution and inducing a change in pH which is detected by the glass electrode [3].

As an example, Figure 2.4 illustrates the functioning mechanism of an NH_3 gas selective probe. Changing the pH of a sample solution containing NH_4^+ ions (1) to pH ~ 9 or higher induces the formation of NH_3 gas. Ammonia can now diffuse across the gas-permeable membrane (2), which changes the pH of the buffer at the other side of the membrane (3). Changes in H^+-activity of this internal solution are sensed by a pH glass electrode (4) and its response is related to the activity of NH_4^+ ions in the original sample solution through the use of standard NH_4^+ solutions.

Similar gas sensor arrangements can be obtained by using a NH_4^+-selective ISE to develop the response to ammonium ions instead of a glass electrode to measure changes in pH, preventing the interference from gases like CO_2 or NO_2 [15,16] which can also generate pH shifts in the internal buffer solution.

2.4.4. Liquid Ion-Exchange Electrodes

2.4.4.1. Cation-Selective

The membrane of these electrodes is a solution of a hydrophobic anion (e.g., tetraphenyl borate) and a cation dissolved in a organic solvent which is immiscible with water [13]. This solution can either impregnate a hydrophobic membrane or be retained in a polymer structure [e.g., poly(vinyl chloride)-(PVC)]. In the presence of an aqueous solution containing the primary ion, the membrane exchanges cations with the solution, and Eq. (2.17) transforms to

$$Ex - l^+_{(mem)} + k^+_{(aq)} \rightleftharpoons Ex - k_{(mem)} + l^+_{(aq)} \qquad (2.20)$$

where Ex is the ion exchanger and k^+ and l^+ are the two cations involved in the process. The selectivity characteristics of these membrane systems are almost exclusively explained by the extraction properties of the solvent used in the membrane. Assuming equal mobilities for similar ionic forms with distribution coefficients k_k and k_l, the selectivity coefficient of Eq. (2.16) can be rewritten as

$$K_{jkl} = \frac{k_l}{k_k} \qquad (2.21)$$

The distribution coefficients depend on the free energy of hydration, $\Delta G^0_{hyd,k^+}$, and solvation of the ions in the membrane, $\Delta G^0_{solv,k^+}$, and its magnitude is given by the equation [10]

$$\ln(k_k) = \frac{\Delta G^0_{hyd,k^+} - \Delta G^0_{solv,\,k^+}}{RT} \qquad (2.22)$$

Because the hydration term is dominant in Eq. (2.22), low free energy of hydration favours inclusion into the membrane. Thus, large organic cations such as acetylcholine, tubocurarin, or lipophilic quaternary ammonium ions (R^+) can be determined in the presence of high levels of inorganic cations in biological applications [10]. The typical selectivity series for cationic species coincides with the Hofmeister lyophilic series:

$$R^+ > Cs^+ > Rb^+ > K^+ > Na^+ > Li^+$$

2.4.4.2. Anion-Selective

Ion-exchange electrodes selective for anions also present a typical selectivity pattern that corresponds to a similar lypophilic or Hofmeister series for anions:

$$R^- > ClO_4^- > I^- > NO_3^- > Br^- > Cl^- > F^-$$

Ion-exchange materials are usually employed in anion-selective electrodes, as synthesizing 3-d cavities (see next entry) formed with positively polarized groups still remains an elusive goal. Hence, electrodes for environmentally important species such as NO_3^- and Cl^- tend to be based on ion exchangers such as quaternary ammonium salts, or inorganic crystalline materials. However, there is increasing interest in the design and synthesis of novel anion-selective receptors whose selectivity departs markedly from the Hoffmeister series, although practical examples are still relatively rare. Recent efforts have focused on the use of positively charged complexes in which the metal center is part of a cavity or area of restricted geometry or shape into which it is hoped the target anion will fit better than common interferents. An excellent review of transition-metal-based receptors for anion-selective recognition and electrochemical–optical sensing has been published by Beer [17]. Other examples include indium(III)-porphryins [18] and organo-tin [19] complexes applied to the determination of Cl^-, palladium organophosphine complex for nitrite [20], and dithiocarbamate complexes of Hg(II) [21].

2.4.5. Neutral-Carrier-Based ISEs

These electrodes are similar to ion-exchange ISEs, but the active species are neutral molecules, C, dissolved in the organic solvent. The corresponding cation complexes formed (k^+C) are electrically charged:

$$k_{(aq)}^+ + C_{(mem)} \rightleftharpoons k^+C_{(mem)} \tag{2.23}$$

These molecules possess a polar cavity with electron donor atoms or negatively polar groups that interact with the ion. The stability of the complex is determined by a "best-fit" mechanism, and the neutral carrier is more selective toward the ion, within the same period, that best fits the molecular cavity. Bigger ions will not be able to access the coordination sites, and the bonding of smaller ions to the coordination atoms will be too weak to form stable complexes. The type of molecules used as neutral carriers is widely variable, ranging from crown-ethers derivatives [22] to calixarenes [23–25] or complex macromolecules and antibiotics like valinomycin [26] or nonactin [27–29].

In general, the molecular structure should comply with the following requirements [10] (see example in Figure 2.5):

1. The molecule must contain both polar and nonpolar groups.
2. The polar groups should form a stable polar cavity suitable for ion coordination. The lipophilic groups should shield the charge of the complexed ion from the membrane solvent (e.g., PVC plasticizer) and ensure sufficient solubility of the complex in the membrane phase.
3. The coordination sphere should present no more than 12 coordination sites, preferably 5–8.
4. Higher selectivities can be obtained if the coordination sphere is fixed to a

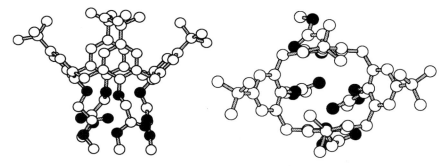

Figure 2.5. Energy-minimized 3D structure of the ionophore Calix[4]:tetrametoxyethyl ester. The picture shows the oxygen atoms that form the negatively charged cavity inside the molecule suitable for ion coordination. Hydrogen atoms are not shown for clarity. (*Key:* ○ carbon; ● oxygen.)

rigid molecular structure. For a periodic series, the cation that best fits the cavity is the one preferred by the ionophore.

5. The ionophore should be small enough to present high mobility, but yet sufficiently large to provide the required preference for a lipophilic medium.

2.5. THE IMPORTANCE OF SELECTIVITY

The Nikolskii–Eisenman equation predicts that the most accurate results will be obtained either when there are no interferents present in the sample or when the effect of the interferents on the electrode response is negligible $\Sigma_l K_{jkl} a_{il}^{z_k/z_l} \Rightarrow 0$). When the effect of the interferents is significant, the electrode response may be completely suppressed, and the device therefore unusable for that application. Hence, during the last 20 years or so, a number of different approaches have been used to attempt to improve the selectivity and broaden the applicability of ISEs.

The following section (2.5.1) deals with improving thermodynamic selectivity, while Section 2.7.2.4 deals with the enhanced selectivity derived from kinetic measurements. If the problem faced is high and variable levels of interferents an ISE array approach may be the only option. Very high selectivity may, in fact, be an obstacle for the use of a particular ISE in a sensor array (see Section 2.10).

2.5.1. Thermodynamics of Selectivity

As described in Section 2.4, the selectivity obtained with electrodes based on ion exchangers depends on the lipophilicity of the ionic species and its distribution coefficient between the membrane and the aqueous phase.

In contrast, neutral-carrier-based ISEs owe their selectivity to the reversible process shown in Eq. (2.23), which can be considered equivalent to the distribution of the ion k^+ in both the membrane and aqueous phase. The equilibrium of that reaction is governed by the constant K_k, which is the overall distribution coefficient:

$$K_k = \beta_{kC} k_k a_C \tag{2.24}$$

This depends on the stability constant of the ion–carrier complex in the membrane solvent β_{kC}, the distribution coefficient (k_k) of the free ion between the membrane and the solution, and the activity of free-carrier molecules a_C in the membrane [10].

To avoid dealing with stability constants in membrane solvent media, the reaction shown in Eq. (2.23) can be studied as a consecutive number of subprocesses:

1. Migration of the free-carrier ligands to the boundary surface with the aqueous solution as in Eq. (2.25). This process is determined by the distribution coefficient of the free ligand between membrane and aqueous phase k_C:

$$C_{(mem)} \overset{1/kc}{\rightleftharpoons} C_{(aq)} \tag{2.25}$$

2. Complexation, in aqueous phase, between the ionic species k^+ and the carrier molecule C [Eq. (2.26)], being β_{kC}^w, the stability constant of the complex formed in the aqueous phase.

$$k^+_{(aq)} + C_{(aq)} \overset{\beta_{kC}^w}{\rightleftharpoons} k^+ C_{(aq)} \tag{2.26}$$

3. Migration of the complex k^+C into the membrane [Eq. (2.27)]. The distribution coefficient of the ion–carrier complex k_{kC} controls this equilibrium.

$$k^+ C_{(aq)} \overset{k_{kC}}{\rightleftharpoons} k^+ C_{(mem)} \tag{2.27}$$

From these equilibria the overall distribution coefficient can be redefined as

$$K_k = \beta_{kC}^w k_{kC} \frac{a_c}{k_c} \tag{2.28}$$

In the frequent case where the electrochemical cell considered contains a second ion of the same charge l^+, which can also permeate the membrane, and assuming equal mobilities for the complexes k^+C and l^+C, Eq. (2.16) can be rewritten for neutral carrier membranes as

$$K_{jkl} = \frac{K_l}{K_k} = \frac{\beta_{lC}^w k_{lC} (a_c/k_c)}{\beta_{kC}^w k_{kC} (a_c/k_c)} = \frac{\beta_{lC}^w k_{lC}}{\beta_{kC}^w k_{kC}} \qquad (2.29)$$

which is consisent with Eq. (2.17) for neutral carrier membranes and quantifies the following competition equilibrium of the parallel complexation reaction with either ionic species:

$$k^+C_{(mem)} + l^+_{(aq)} \underset{\phantom{K_{jkl}}}{\overset{K_{jkl}}{\rightleftharpoons}} l^+C_{(mem)} + k^+_{(aq)} \qquad (2.30)$$

Improvement in the selectivity of an ISE membrane may therefore be obtained through modification of the neutral carrier molecules (β_{kC}^w) and/or the composition of the final membrane cocktail (k_{kC}).

Equation (2.29) clearly shows the dependence of the potentiometric selectivity of a neutral-carrier-based electrode on the stability constant of the complexes formed between the carrier ligand and the ions, and that the relative selectivity for two different ions is proportional to the ratio of these stability constants β^w, provided that the distribution coefficients are similar.

2.5.2. Optimization of Membrane Composition

In general, these selective membranes consist of the neutral carrier or ionophore, the polymer, usually poly(vinyl chloride) (PVC), and a plasticiser which acts as the organic solvent. The role of the PVC matrix is to provide an inert solid support structure in which the rest of the components are embodied. The relative proportions of the components affects membrane parameters such as electrode slope, selectivity against interferents, and active life. Different plasticizers, including *ortho*-nitrophenyl octyl ether (*o*-NPOE), di(2-ethylhexyl) sebacate (DOS) or di(2-ethylexyl) adipate (DOA), will affect the lipophilicity of the PVC membrane, which, in turn, alters the distribution coefficients (k) of the different species. The more hydrophilic plasticisers, like *o*-NPOE, will have higher distribution coefficients and hence higher ion activity inside the membrane [Eq. (2.24)], but the ligand may leach out to the aqueous phase more easily, reducing the working life of the electrode. In general, the more hydrophilic plasticizers will also alter the M^{2+}/M^+ selectivity ratio, favoring the M^{2+} species because of the increased polarity of the membrane [10].

Other additives may improve the membrane characteristics, including potassium tetrakis(4-chlorophenyl)borate (KTpClPB), which functions as an anion excluder, preventing negatively charged ions entering the polymer membrane and helping to generate a more ideal Nernstian response. However, a minimum amount should be used as the exchanger anions have their own cation exchanger selectivity which typically favors more larger, more lipophilic cations as there is no best-fit mechanism and this will usually conflict with the desired selectivity.

2.6. ANALYSIS WITH SINGLE ISEs

Potentiometric measurements must be made at zero current. Previous to the 1960s, large, high-power voltmeters based on valve transistors were used which, although accurate, were not portable and required considerable warmup periods. Since the 1960s, operational amplifier (op-amp)-based instruments with very high input impedance (typically ~ 1 TΩ) have been developed. These devices allow potentiometric measurements under near-zero-current conditions to be made using small low-power, lightweight integrated circuits.

Potentiometric devices respond in a logarithmic fashion, as the change in potential is related to the logarithm of the ion activity [Eq. (2.3)] with a maximum theoretical sensitivity (in the best case of ion charge being $+1$) of 59.16 mV for a 10-fold change in ion activity at 25°C. This means that, for a minimum error of 1%, the instrument used must have a discrimination factor of at least 0.25 mV.

2.6.1. Important ISE Characteristics

A number of characteristics are required from an ISE so it can be consider a suitable sensor for quantitative ion analysis. Of these, slope, selectivity, limit of detection, and electrode working life are probably the most important.

2.6.1.1. Slope

This is a measure of the sensitivity of the electrode, and the closer it is to the Nernstian value of $59.16/z_k$ mV per decade at 25°C [Eq. (2.3)], the more ideal the electrode behavior is. In routine determinations with ISEs the electrode working range will be determined by the activity window of the primary ion over which the electrode response is Nernstian (Figure 2.3).

2.6.1.2. Limit of Detection

The limit of detection (LOD) can be used to compare the performance of different electrodes at low primary ion activities. One definition describes the LOD as the value of primary ion activity where the activity of the primary ion equals the summation term for interferents in the Nikolskii–Eisenman equation (Figure 2.3), specifically, the contribution of all the interferents equals the contribution of the primary ion [10]. At this point[2]

---

[2]Equation (2.31) can also be used to calculate the selectivity coefficient for a particular interesting ion l if a_{ik} and a_{il} are known.

$$a_{ik} = \sum_l K_{jkl} a_{il}^{z_k/z_l} \tag{2.31}$$

Comparing the difference in response from primary ion only (E_{ik}) to mixed behavior (E_{ikl}), we obtain

$$\Delta E_i = E_{ikl} - E_{ik} = S_j \log\left(a_{ik} + \sum_l K_{jkl} a_{il}^{z_k/z_l}\right) - S_j \log a_{ik} \tag{2.32}$$

$$\Delta E_i = S_j \log \frac{2a_{ik}}{a_{ik}} = S_j \log 2$$

This means that the LOD for an electrode selective for a single charged ion is the activity level corresponding to where the mixed response deviates from the primary response by 17.8 mV ($17.8/z_k$ for a z charged ion). Limit of detection does not mean limit of determination, and ISEs can be used for analysis of activities below the LOD using nonlinear calibration approaches, although it is obviously preferable to work in the linear Nernstian range.

2.6.1.3. Selectivity

Selectivity coefficients give an idea of the degree of discrimination of an electrode to ions in solution other than the primary ion (see also Section 2.5.1). The smaller the coefficient, the better the selectivity of the ISE. However, values are far from constant and depend on the activity range studied and the measurement technique used [30,31].

As an example, if the overestimation accepted for the total primary ion concentration is a maximum of 5%, then, for binary mixtures where primary (k) and interfering (l) ions ($z_k = z_l$) are present in the solution at the same level, the electrode used should have a selectivity coefficient $K_{jkl} = 5 \times 10^{-2}$ or lower.

2.6.1.4. Electrode Lifetime

The working life of an ISE can vary from a few days to a few months, and will depend on factors such as the analysis technique used, the matrix of the samples analyzed and the membrane composition. The lipophilicity of the plasticiser used in the membrane is a major factor affecting the life time of PVC membrane electrodes [10]. With these sensors, deterioration of the selective species, or leaching out of one or more components into the aqueous samples will produce a gradual decrease of the electrode slope and eventually null response. An important parameter to consider during the active life of the electrode is how long the ISE can be used between calibrations without significant changes in the electrode parameters that will affect the accurate prediction of unknowns.

2.6.2. Sources of Error

2.6.2.1. Activity

As explained in Section 2.2, the response of an ISE is related to the activity, not concentration, of the ions in solution. The analyst must always bear in mind what the relationship between activity and concentration is and how this varies from solution to solution. This relationship is defined as

$$a_k = \gamma_k C_k \tag{2.33}$$

where γ_k is the activity coefficient of the ionic species k in a particular solution. Several theories exist to explain this relationship, all of them relating the values of the activity coefficient to the ionic strength of the solution. The Davies equation, which is widely used, states that this relationship is expressed by

$$\log\gamma_k = -A\left(\frac{z_k^2\sqrt{I}}{1 + \sqrt{I}} + cI\right) \tag{2.34}$$

where z_k is the ionic charge of the species, A and c are parameters that take the values of 0.51 and 0.15 for aqueous solutions at 25°C [32], and I is the ionic strength of the solution, which is defined in terms of all the ionic species k present in the sample:

$$I = \frac{1}{2}\sum_k z_k^2 C_k \tag{2.35}$$

The Davies equation is an extended form of the Debye–Hückel equation with the advantage that it can be applied for ionic strengths up to 0.6 mol/dm^3 (see Figure 2.6) [33].

The activity of the primary ion must, of course, be high enough to be within the dynamic range of the ISE. The minimum activity that can be accurately analysed will depend on the type and activities of interferents present in the sample and the electrode selectivity. Indifferent ions that do not induce a direct response in the ISE can nevertheless affect the electrode potential by changing the ionic strength of the solution and therefore the activity coefficient of the primary ion (equal concentrations of primary ion do not automatically lead to equal potentials!). Furthermore, formation of complexes between the primary ion and ligands in the sample will lead to underestimation of total ion concentration, as this will reduce the free-ion activity, to which these electrodes respond. On the other hand, if the analyst is interested in the free ion activity rather than total concentration (as often is the case

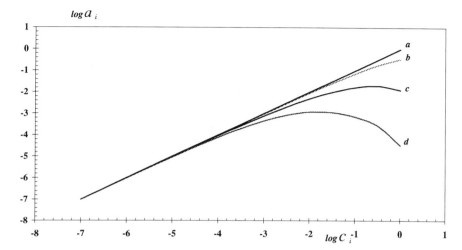

Figure 2.6. Effect of the ionic strength on the activity of single-ion solutions. Deviation from the ideal relationship (*a*) for a single-charged ion (*b*), for a double-charged ion, (*c*) and a triple-charged ion (*d*) as calculated from the Davies formalism [Eq. (2.34)].

in physiology and biochemistry studies) ISEs may be the only suitable analytical method.

2.6.2.2. Interferents

The contribution of the interferents to the ISE signal should, ideally, be much smaller than that of the target analyte, in which case they can usually be neglected. If the matrix is known, then it is important to match as closely as possible the background matrix composition of the samples and calibrants in order to minimize variations in ionic strength, complexation, and junction potential. If the matrix composition is unknown and variable, analysis of the target analyte can be performed on a sample to sample basis using a standard addition procedure (Section 2.6.4.2) or sensor array approach (Section 2.9).

2.6.2.3. Matrix Effects

The presence in the sample matrix of other species than interferents, such as proteins or lipids, may create suspensions or gels of high-molecular-weight particles that can affect the proper performance of the ISE by deteriorating the membrane or dissolving its components, therefore considerably reducing its working life. In addition, the presence of macromolecules such as proteins or long chain lipids may coat the membrane and eventually block the active electrode surface. In flowing systems this same effect may produce clogging of narrow flow conduits.

In addition, chelates or other complexing agents may favor the development of side reactions that will interfere the proper determination of the analyte by reducing the level of activity of the free ion. Furthermore, the matrix composition and ionic strength will affect the reference electrode junction potential. Changes in the junction potential will generate an error in the analytical measurements, and therefore the matrix composition and ionic strength should be matched as closely as possible in samples and standards.

2.6.3. Calibration and Characterization

In batch analysis with ISEs, the electromotive force represents the potential of the whole of the electrochemical cell and, assuming the potential of the reference electrode (E_{ref}) remains constant, the variations in the cell potential from solution to solution can be related to variations on the potential of the ISE alone [$E_{(ISE)}$]:

$$E_{(cell)} = E_{(ISE)} - E_{(ref)} \qquad (2.36)$$

The general characteristics of an ISE are usually obtained by studying the steady-state response to a wide range of primary ion activity, typically between 10^{-7} and 10^{-1} mol/dm^3. This calibration gives general electrode parameters such as the standard electrode potential, slope, and LOD. Observing the response to solutions containing interfering ions will provide information about the selectivity coefficients. If very precise measurements are needed in a particular application, a different calibration set more focused on the sample range can be designed, giving information on how the electrode behaves in that specific activity range. For best results, constant temperature conditions should be employed, as factors such as electrode slope [see Eq. (2.2)], activity coefficients, and standard cell potentials are temperature-dependent.

It is important to know the effect of ionic strength on the activity of the ions in the standards. This effect rapidly increases in significance with concentration and is much greater for multiply charged ions. The relationship between activity (a_i) and concentration (C_i) predicted by the Davies equation [Eq. (2.34)] is shown in Figure 2.6. For salts of singly charged ions (M^+X^-) (b), the deviation from the ideal relationship (a) becomes apparent only when the concentration becomes greater than $\sim 10^{-2}$ mol/dm^3. This deviation is more dramatic for double-charged (c) and triple-charged ions (d). Modifying the calibration solutions so that they contain the same high concentration of indifferent ions gives a relatively constant ionic strength over the whole calibration range and therefore effectively constant activity coefficients. Electrode potentials or potential changes can then be directly related to the ion concentration, but sometimes at a cost of reduced electrode sensitivity, as the slope may decrease because of a reduction of the activity of the primary ions arising from dilution effects. A key task is therefore to identify an ionic solution that has no direct effect on the ISE or reference electrode potentials so that it can be added at higher concentrations to samples and standards in order to smooth out any differences in ionic strength as, in this situation, the overall ionic strength is dominated by the indifferent ions added [Eq. (2.35)]. The solution used for this purpose is known as *ionic-strength adjuster* (ISA).

For any two solutions (1) and (2) of primary ion activity a_1 and a_2:

$$\Delta E = S_j \log\left(\frac{a_1}{a_2}\right) = S_j \log\left(\frac{c_1 \gamma_i}{c_2 \gamma_2}\right) \tag{2.37}$$

If the ionic strength remains constant, then $\gamma_1 = \gamma_2$ and therefore the ISE response can be related to the concentration of the ion in the sample:

$$\Delta E = S_j \log\left(\frac{c_1}{c_2}\right) \tag{2.38}$$

2.6.4. Analysis and Prediction

A number of analytical approaches are typically employed with ISEs, which vary in complexity and performance. In brief, these can be summarized as follows.

2.6.4.1. Direct Potentiometry

After pre-treatment and electrode calibration, direct potentiometry involves presenting the sample to the ISE and reference electrode, and fitting the change in the cell potential to the calibration parameters using a graphical approach (Figure 2.3) or increasingly, by means of parameters (E^0_{cell}, S, etc.) and algorithms stored electronically within the instrument. These methods require that the matrix of the calibration solutions and the samples are matched, as the only variable allowed is the activity of the primary ion. This is ideal for samples for which the contribution to the electrode signal by interferents is negligible.

2.6.4.2. Standard Addition

This technique is usually employed when the sample matrix has an unknown but probably significant effect on the cell potential. First, the cell potential in the sample is recorded and then a volume of a solution containing a known concentration of the primary ion is added. This is commonly a small volume of concentrated primary ion solution to minimize dilution effects. The cell potential in this altered sample is also recorded. The change in response is related only to the change in the activity of the primary ion, and from the Nernst equation the difference in electrode potential can be described as

$$\Delta E_{ij} = E'_{ij} - E_{ij} = S_j \log\left(\frac{a'_{ik}}{a_{ik}}\right)$$ (2.39)

where E'_{ij} and a'_{ik} are the electrode response and activities after the primary ion addition. From Eq. (2.39) the initial primary ion activity in the sample can be determined from the equation

$$a_{ik} = \frac{V_{std}a_{std}}{V_i + V_{STD}}\left[10^{\Delta E_{ij}/S_j} - \left(\frac{V_i}{V_i + V_{std}}\right)\right]^{-1}$$ (2.40)

where a_{std} and V_{std} are the activity and the volume of the standard solution added and V_i is the initial volume of the sample. If a relatively small aliquot ($V_{std} \Rightarrow 0$) of high primary ion concentration standard solution is added, the volume can be considered constant and the activity of the primary ion in the original sample is given by the following simplified equation:

$$a_{ik} = \frac{V_{std}a_{std}}{V_i}\left(10^{\Delta E_{ij}/S_j} - 1\right)^{-1}$$ (2.41)

In order to use this approach, the electrode slope (S_j) must first be measured by using two standard solutions. Then the standard addition is performed.

A number of additions to the same sample can be used to generate a Gran's plot, increasing the accuracy of the technique with the number of spikes used on the sample. Assuming that there is no change in ionic strength and that the junction potential remains constant, the measurements from each of the additions can be used to construct a graph similar to Figure 2.7 containing the response [as $a_{ik}(V_i + V_{std})10^{\Delta E_{ij}/S_j}$] versus volume of standard added (V_{std}). Rearranging Eq. (2.20), we obtain

$$a_{ik}(V_i + V_{std})10^{\Delta E_{ij}/S_j} - a_{ik}V_i = a_{std}V_{std}$$ (2.42)

Thus, extrapolating the graph until it crosses the x axis provides a point where $a_{ik}(V_i + V_{std})10^{\Delta E_{ij}/S_j} = 0$, and the activity of the original sample can be calculated as:[3]

---

[3]Note that, because of the value of V_{std} obtained from the graph is negative, the equation gives a positive value of the original sample activity.

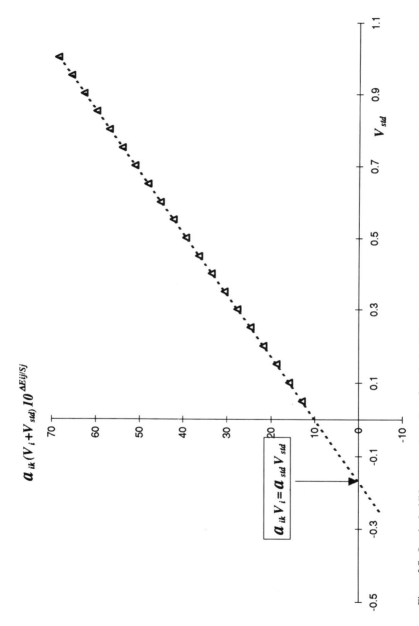

$a_{ik}(V_i + V_{std})10^{\Delta E_{ij}/S_j}$

$a_{ik}V_i = a_{std}V_{std}$

V_{std}

Figure 2.7. Standard additions curve. A number of standard addition procedures can be applied to the same sample to increase the accuracy in the determination of the original sample concentration.

41

$$a_{ik} = \frac{a_{std}V_{std}}{V_i} \qquad (2.43)$$

Alternatively, the original sample activity can also be obtained from the slope of Gran's plot.

2.6.4.3. Titrations

Ion-selective electrodes can also be used as potentiometric equivalence point indicators in titrations. Monitoring the change in electrode potential with the addition of a known volume of titrant gives a measure of the change in concentration of a particular species during the titration procedure. By far, the most important commercial application in this respect is the routine determination of pH using dedicated autotitrators.

In automated acid–base titrations, the change in potential of a glass electrode is used to identify the equivalence point of the titration. Figure 2.8 illustrates the variation of the pH of an ethanoic acid solution with the addition of a known volume of standardized NaOH solution. The equivalence point location, which in this particular example corresponds to an addition of 10 mL of titrant, can be easily

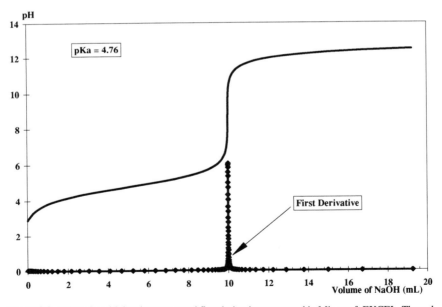

Figure 2.8. Ethanoic acid titration curve and first derivative generated in Microsoft EXCEL. The volume of titrant base needed to reach the equivalence point is 10 mL.

identified from the pH profile obtained when an excess of titrant (twice as much as necessary to reach the equivalence point) is added to the initial solution. Although in this particular case the equivalence point is quite clear, weak acids or bases, and in particular polyprotic species, may present titration profiles with diffuse or nonvisible equivalence points. In these situations, the first or second derivative of the pH trace can be used to accurately identify the precise volume of titrant used (See Figure 2.8) for each group neutralized.

Alternatively, Gran's plots can be employed to accurately predict the volume needed to reach the equivalence point of a titration without the need for complete reaction. This is achieved by plotting $V_b 10^{-pH}$ against V_b for a number of measurements obtained before the equivalence point, where V_b is the volume of titrant added to the sample. The straight line obtained should present a slope equal in value to the acidity constant of the species ($S = -K_a$). Extrapolating these data to the point where the line crosses the x axes will give the estimated value for the volume of titrant needed to reach the equivalence point. Regression analysis on these data will also give an estimate of the quality of the result through the correlation coefficient (r) of the points used for the prediction and the standard deviation of the x-axis intercept. Figure 2.9 shows how this method can be applied to the prediction of the equivalence point in the titration of ethanoic acid (Figure 2.8).

ISEs selective for other ions can be used in autotitrators for monitoring the variation of the target species in other chemical systems such as complexation equilibria or redox (reduction–oxidation) processes.

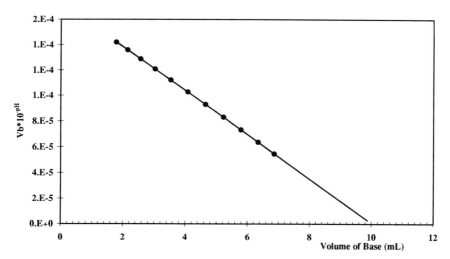

Figure 2.9. Gran's plots of ethanoic acid titration. Gran's plots can be used to accurately predict the volume of titrant needed without the need to neutralize completely the species being titrated. V-axis intercept when $V_b = V_e$. Slope of regression line $= -K_a$ $y = -1.72E - 05x + 1.73E - 04$.

2.7. FLOW-INJECTION POTENTIOMETRY

Ion-selective electrodes can be used in both batch measurements and flow-injection analysis (FIA). Although the chemical composition of membranes do not differ significantly, factors such as thickness, the type and architecture of the electrode body, and the membrane position can affect dramatically the overall response behaviour in FIA. Thus, membranes from the same production cocktail used in batch and flow analysis will show very different response characteristics and different electrode parameters in FIA and batch measurements. Also, different possible flowing geometries in FIA affect the ISE response, which indicates that the overall detector characteristics do not solely depend on the analysis technique, but also on the way the sample is presented to the electrode membrane.

2.7.1. Electrode Geometries

Figure 2.10 shows the most common electrode geometries employed in batch and flow-injection measurements. The typical design for a bench-type electrode and its components is shown in Figure 2.10a. An internal reference electrode, usually Ag/AgCl, is immersed in a solution containing a fixed concentration of the primary ion (typically 0.1 mol/dm^3). This solution is in contact with the internal surface of a selective membrane, creating a constant potential at the inner boundary. The electrode body is immersed in the sample solution and a potential difference, which depends on the activity of ions in the solution, is created across the membrane.

Figure 2.10b shows the most common designs for flow-injection ISEs [34]. In the flow-through (i) and flow-past geometries (ii) the sample crosses the electrode

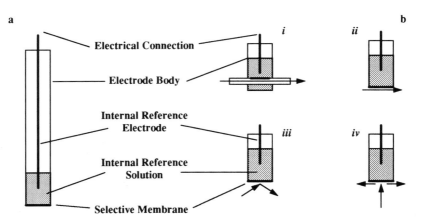

Figure 2.10. Ion-selective electrode geometries: (a) Bench or dip-type electrode for batch measurements; (b) ISE geometries for flow systems; (i) flow-through; (ii) flow-past; (iii) tangential; (iv) wall-jet.

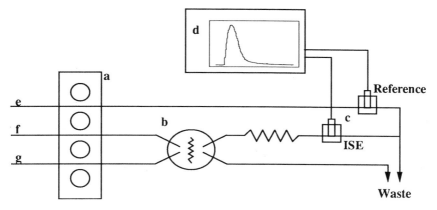

Figure 2.11. Setup diagram for the FIA system with single ISE detector. The minimum setup for an FIA system must consist of the following parts: (*a*) pump; (*b*) injection port; (*c*) detector—ISE and reference electrode; (*d*) measuring device. The different stream lines carry: (*e*) reference electrode internal electrolyte; (*f*) carrier; (*g*) sample.

body through a small conduit in which part (or the whole) of the wall has been replaced with the selective membrane [35,36]. In the tangential electrode (iii) the sample strikes the membrane at an angle between 0° and 90°. Inflow and outflow angles usually have the same value. In type iv, or wall-jet, the sample reaches the membrane at an angle of 90° and exits at the sides [37].

In a typical FIA setup, the ISE is placed in the carrier stream, and the injector diverts a sample plug into the carrier stream (Figure 2.11). The reference electrode placed in a parallel stream, usually KCl, which is pumped continuously at a slow rate and forms a very reproducible junction with the carrier downstream from the ISE.

2.7.2. Advantages of Flow-Injection Potentiometry

2.7.2.1. Drift

The traditional batch analysis approach involves dipping a bench-type ISE and a reference electrode into the solutions and reading the electrochemical cell potential after the steady state is allowed to develop (typically 1–3 mins). This is to assure that the full electrode response to the primary ion is more or less obtained. In batch measurements, the cell potential obtained is related to a calibration curve or to a electrode response equation from which unknowns can be estimated. Electrode potential drift constitutes a serious problem, as it cannot be compensated for unless the ISE is regularly recalibrated.

In contrast, with FIA systems employing ISEs as detectors the analytical result is obtained from the change in potential obtained when a sample plug injected into the carrier stream passes by the electrode membrane. Usually the carrier contains a low concentration of the analyte ion in order to stabilize the ISE membrane potential, and the sample plug contains a higher concentration of analyte. Hence the analytical signal is typically in the form on an unsymmetric peak [43–45] with a tailing end reflecting both the effect of hydrodynamics on the sample plug, and the slower kinetics of ion release compared to ion uptake exhibited by these membranes (see Figure 2.12). An important advantage of FIA therefore is the continuous reference signal provided by the carrier between measurements, unlike batch measurements, where the reference signal might be checked only every hour or so. In addition, as the analytical measurement is usually the peak height, any background drift is automatically compensated for, as the baseline is subtracted from the peak maximum to obtain the analytically important quantity. This leads directly to improved accuracy in the analytical results.

2.7.2.2. Reproducibility

Because of the highly reproducible hydrodynamics of sample transport and presentation to the ISE membrane from the injector port, the concentration profile of the sample plug is very reproducible. The signal is developed and decays in the same manner from sample to sample, leading to excellent precision in the results. The washing effect of the carrier also ensures that sample carryover is negligible. These points are nicely illustrated in Figure 2.13. Furthermore, this reproducibility implies that multiple measurements can be made with a single FIA peak, not just at the peak maximum (although with computerized systems, individual points in the peak maximum can be used either as separate analytical results or averaged to produce a more precise measurement). Better linearity and selectivity may in fact be obtained from points away from the peak maximum. This situation is somewhat analogous to the processing of UV–vis spectra obtained from photodiode array spectrometers, where any portion of the spectrum can be used to generate the analytical result, not just peak maxima. The ability to use any part of the response for analytical purposes makes FIA particularly suitable for kinetic studies [38,39] in situations where the sample has a suitable chemical interaction with the sensor membrane. This is thoroughly exploited in some of the more advanced FIA setups like double injection [40], stopped-flow [45], or reverse-flow [41] systems.

2.7.2.3. Sample Turnaround

Compared to other flowing techniques that use continuous or air-segmented flows, the speed of the carrier along the FIA conduits is very high. The flow rate used is typically between 0.5 and 1.5 mL/min or even higher. This means that at a flow rate of 1.0 mL/min (typical), a sample volume of 100 μL stays in contact with the

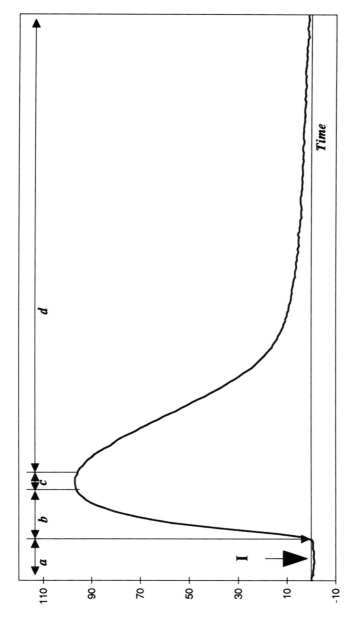

Figure 2.12. Typical profile of the transient ISE potential in FIA systems. When the electrode is in equilibrium with the carrier solution (*a*) and a sample plug is injected (*I*), the ISE potential presents a very rapid increase due to the boundary potential followed by a decrease in the rate of potential development (*b*). A *pseudo*-steady state is reached at the top of the FIA peak (*c*). As the sample plug leaves the detector and the carrier solution washes the detector, the electrode potential returns to the baseline (*d*).

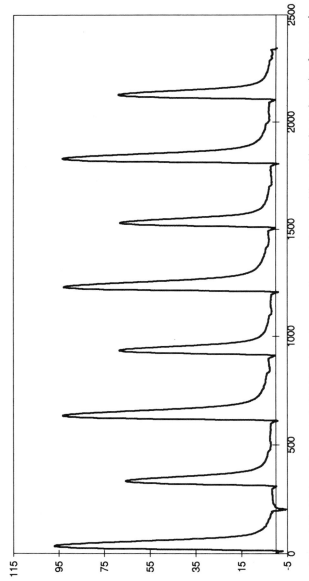

Figure 2.13. FIA traces from a single ISE. The electrode response is generated very rapidly, reaching the maximum in a few seconds. Return to baseline takes longer, due to the slower kinetics of ion release from the membrane, but usually within 1–2 min after an injection the potential has returned to the baseline. The system is now ready for the next sample injection. In this figure, the response to two alternating standards of a PVC membrane sodium electrode clearly shows excellent reproducibility and no sample carryover effects. (x-axis scale is in data points, sampling rate 5 Hz.)

detector for only 6 s, during which the response maximum is developed related to the sample concentration. Provided there is a fast return to baseline, ISE-FIA systems can produce sample throughputs of 120 samples per hour or even higher. Coupled with the ease of automation, this makes the technique very attractive for many on-line monitoring situations in industry. In computerized systems, the user can exercise complete control over the system parameters such as carrier flow rate, sample injection volume, and injection frequency.

2.7.2.4. Kinetic Selectivity

When a membrane in equilibrium with the primary ion is exposed to a solution of higher primary ion activity, a fast initial change in membrane potential occurs (boundary potential), followed by a slower potential settling toward the steady-state plateau, which can typically take between 1 and 3 min (bulk-membrane diffusion potential). The first effect is kinetically limited and is mainly a response caused by the primary ion due to the faster rate of exchange of these ions compared to interferents, which gives some degree of kinetic enhanced selectivity [42]. This fast initial response depends on how the sample is presented to the membrane (e.g., stirring rate), and, as it is not very reproducible in batch measurements, it cannot be exploited for analysis purposes using that experimental approach.

On the other hand, when the potential settles, the exchange process is thermodynamically controlled (Figure 2.14). The equilibrium involves a competition mechanism between the primary ion and interfering ions to complex with free ligands, and therefore the exchange selectivity in steady-state will depend on the stability constants of the ion–ligand complexes formed in the membrane (see Section 2.5.1). This steady-state potential E_{ss} is the only part of the ISE response in the batch regime reproducible enough to be related to the primary ion activity in the solution. However, in flow-injection systems [43,44], the sample is transported to the electrode active surface through tubes of small diameter. The sample occupies only a small section of the flow between two sections of a carrier or conditioning solution, which pushes the sample to the detector. When the electrode membrane comes in contact with the sample plug, a potential (E_{FIA}) is developed (typically in less than 10 s)[4] relative to that generated by the carrier solution. The profile of the peak response is very reproducible, and therefore, in principle, one is not restricted to the peak maximum for analytical measurements, as any point on the profile can be related to the analyte activity [45].

Figure 2.15 simulates responses for batch and FIA measurements with the same electrode to separate solutions containing equal activities of a primary ion (k) and an interfering ion (l), respectively, for a sitatuation where both ions have the same

---

[4]In this situation, the potential developed may not reach the steady-state and a sub-Nernstian response often occurs. See Figure 2.14.

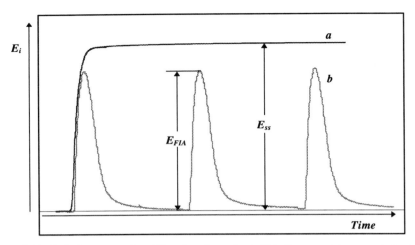

Figure 2.14. Typical response of an ISE to the primary ion in FIA. Response (*a*) represents the development of the response of ISEs in batch measurements. The potential generated E_{ss} is maximum when the steady-state equilibrium is reached. For flow-injection measurements (*b*), the concentration is related to the potential E_{FIA} observed at the maximum of the transient electrode response, which is a function of the contact time between the sample plug and the membrane surface as well as the ion-exchange kinetics and thermodynamics.

thermodynamic behavior, but differing kinetic behavior. Hence, in batch measurements, the same steady-state potential is obtained in both cases (E_{ss}), but the primary ion (a_k) reaches the steady state more rapidly than does the interfering ion (a_l). In contrast, in FIA (compare b_k and b_l), using a fixed contact time between the sample plug and the membrane that falls within the kinetic dominated region of the membrane potential development, there is a marked difference in the peak heights, with the primary ion peak ($E_{k,FIA}$) much larger than the corresponding interfering ion peak ($E_{l,FIA}$). As the kinetics of uptake and release generally favors the primary ion, there clearly exists a major opportunity to exploit kinetic selectivity [46,47] by optimising the system parameters that control the timebase of the peak (injection volume and flow rate).

2.7.3. Reagent Consumption

The volume of sample injected into the carrier stream will vary depending on the application but is usually between 25 and 200 μL. This volume is much smaller than the milliliter scale, which is usually the case in batch analysis.

For applications where the sample is available in very small quantities, it can be injected into the carrier stream after additon of reagents or dilution (if required). In the opposite case, where the sample is inexpensive and available in large quan-

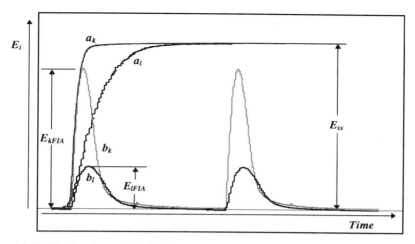

Figure 2.15. Kinetic-enhanced selectivity in ISEs. In flow-injection systems, the difference in electrode response time to the primary ion and the interferents can significantly increase the selectivity for the primary ion in comparison to batch techniques (see text).

tities, for instance, in seawater, standards can be injected in very small volumes into the sample, reducing the amount of reagent needed for the analysis.

2.7.4. Sensor Lifespan

As seen above, the proportion of analysis time the electrode membrane is in actual contact with the sample can be very short, and between samples, it is washed by the buffer or conditioning solution used as carrier. In those cases where the target analyte is present in a sample matrix that may contain species capable of damaging or reducing the sensor's ability to detect the analyte, a FIA approach can considerably increase the working life of the detector.

On the other hand, membrane components with relatively high hydrophilicity may leach out of the membrane phase, being washed away in the carrier. This will inevitably cause a faster than usual deterioration of the electrodes as the membrane components will never reach steady-state concentration in the carrier phase.

2.7.5. Sensor Conditioning

Because the selectivity mechanism involves ion complexation in the boundary region between the membrane and the aqueous solution, the electrodes must be conditioned regularly to ensure that the membrane–solution interface is stable. In conventional batch analysis this conditioning must be done after a small number of measurements (if not after every single one). This is obviously very time consuming and it can increase considerably the experimental time for sample analysis.

On the other hand, in FIA systems, the carrier solution is continuously washing through the detector between analysis, which helps to stabilize the boundary potential in the interface region. If the buffer contains a low concentration of the primary ion, an exchange equilibrium is established between the membrane and the solution, and once the initial equilibrium is established, the membrane responds faster to higher concentrations of primary ions in the sample.

2.8. APPLICATIONS OF ISEs

Numerous publications on neutral-carrier- and ion-exchanger-based ISEs and their applications can be found in specialized publications as part of the general area of chemical sensors [4–6] or as a class of its own [7]. Other reviews discuss the use of potentiometric selective electrodes for water analysis [48] and biochemical applications [8]. General guidelines for the use of ISEs in environmental applications [49] and as detectors for flow systems [50] have also been reported.

Hundreds of articles on particular applications of ISEs are available in the literature, but, because this section is not intended as an exhaustive compilation of ISE applications and reviews, only general reviews have been considered, and the following sections are compiled mainly from these referenced articles.

Table 2.1 summarizes the receptors usually employed for the selective detection of the commonly occurring inorganic cations and anions with membrane-based potentiometric sensors. There are many more references in the literature to species like organic ions, proteins and enzymes, vitamins, drugs, medicines, and other species of complex molecular composition that can also be detected with ISEs.

Apart from the traditional batch mode for single-electrode potentiometry, ISEs are commonly used as detectors in autotitrators, where the electrodes are used to detect the endpoints of the titration. Industrially, the use of glass pH electrodes in autotitrators is, by far, the most important application (see Section 2.6.4.3).

Most of the flow analysis techniques can benefit from the wide variety of available ISEs and target species. In chromatography, ISEs are not often applied except, maybe, for ion detection after column separation [51]. Air-segmented continuous analyzers and flow-injection systems routinely use ISEs as electrochemical detectors, particularly for the routine measurements of blood electrolytes in hospitals. Capillary electrophoresis has developed as an important separation technique for a wide variety of charged species, and ISEs have been employed as detectors. However, because of the small size requirements for these instruments, solid-state electrode designs such as coated wires are preferred. Examples include

Table 2.1. Active Species for Cation-Selective Membranes[a]

Inorganic Cations

H^+	N, N-dioctylaniline and tridecylamine	Cu(II)	Dithiocarbamates, thiuram sulfides
Ammonium	Nonactin, narasin, monensin, salinomycin	Ag^+	Monothia–crown ethers
Li^+	1, 10-Phenantrolines, crown ethers, formazans, isatin oximates	Ni(II)	Bis(2-ethylexyl)phosphate
Na^+	Calix[4]arene esters and amides, crown ethers, cyclodextrins	Zn(II)	Tetradecylammonium tetrathiocyanatozincate
K^+	Pyridine macrocycle, valinomycin, bis-crown ethers	Pb(II)	Tetraphenylborate surfactant complexes
Cs^{2+}	Calix[6]arene derivatives	Fe(III)	1,7-Dithia-12-crown-4
Ca^+	ETH series, lipophilic hexapeptides	Tl(III)	Butylrhodamin β-tetrachlorothallate
Mg^{2+}	Acyclic aspartamides	Au(III)	Butylrhodamin β- and tetraphenylpyridinium tetrachloroaurates
Ba^{2+}	Barium (tetraphenyl borate–Antarox CO-880 complex)	U(VI)	Uranyl bis(phosphates)

Inorganic Anions

Clhloride	Silver chloride–silver sulfide	Perchlorate	Berberine hydrochloride, N-ethylbenzo thiazole-2,2´-azaviolene perchlorate
Iodide	Vitamin B_{12} derivatives	Phosphate	Bis(p-chlorobenzyl)tin dichloride
Nitrite	Tetraalkyitin compounds	Thiocyanate	Nitron-thiocyanate complex
Sulfite	Bis(diethyldithiocarbamato) mercury(II)	Chlorocromate	Quaternary ammonium chlorocromate
Carbonate	Ion-exchanger-derivatized benzoic acid ester complex	Periodate	Berberine periodate

[a]Columns on the left indicate the target analyte; on the right, the active materials for the selective electrodes [4–7].

the detection of weak aliphatic acids [52] and chloride, nitrate, nitrite, bromide, iodide, and perchlorate [53].

ISEs are increasingly being applied in process situations where real-time information is required. Table 2.2 gives a summary of some of these, with references for further reading.

**Table 2.2. Review of Processes Utilizing Membrane
ISEs as Potentiometric Detectors**

Fundamental studies	Environmental
Determination of ion conductance	Analysis of effluents
Determination of stability constants	Determination of contaminants in wastewater streams
Determination of diffusion coefficients of ions in soil	Monitoring of nuclear waste containment
	Analysis of pH in rainwater
Determination of kinetic parameters of enzymatic processes	Analysis of drinking water
	Soil analysis
Industrial processes	
Pharmaceutical analysis	Fundamental analysis
Fermentation control	Multicomponent analysis
Determination of boric acid in plating baths	Analysis of organofluoride compounds
Measurement of Na^+ and K^+ in wines	Determination of trace-level impurities in nuclear materials
Determination of the chloride content of fresh concrete	Potentiometric titrations
	Enantiomer selectivity
Biochemical applications	
Microsampling of biochemical systems	Clinical analysis
Antigen and antibody monitoring	Analysis of Na^+, K^+, Li^+, Ca^{2+}, CO_2, amines in
Analysis of ions inside cells	Saliva
	Tissue
Determination of H_2O_2 from biological processes	Amniotic fluid
	Whole blood
	Plasma
Petroleum industry	Urine

2.9. RECENT TRENDS AND DEVELOPMENTS

The development of novel ion-sensing technology continues to be a dynamic area for research, with the main trends as follows:

- The use of sensor arrays to build up "intelligent" sensing systems that can monitor and interpret changes in their environment and advise the user of possible malfunctions. This area has been stimulated by the availability of cheap, powerful computers and data acquisition cards with sophisticated software development environments (e.g., Labview from National Instruments, Austin, TX, USA).

- Investigations on the design and fabrication of novel solid-state potentiometric sensor configurations—this area combines research into new polymeric substrates for generating membranes and films that are compatible with planar electrodes, strategies for immobilization of membrane components ranging from entrapment to covalent bonding, and new techniques for generating electrodes (e.g. screen printing, vapor deposition of materials).

- The development of optically responsive ion-selective membranes with properties similar to their porentiometric counterparts. This research has been stimulated by rapid improvements in optical instrumentation [cheap, powerful PCs and data acquisition cards, miniature photodiode array-based UV–vis spectrometers (e.g., the PS1000 from Ocean Optics), and fiber-optic/planar waveguide probes].

These areas will be discussed in detail in the following sections.

2.10.   ISE ARRAYS

Analysis with a single ISE depends on extreme selectivity if the signal obtained is to be interpreted as being overwhelmingly dependent on the primary ion. The use of a sensor array approach is attractive for several reasons, including the provision of multicomponent data (one analyte per sensor in the array) at no cost in turnaround, and the possibility of using pattern recognition techniques to interpret the combined array response, rather than treating each signal as a separate entity. This raises the possibility of building "intelligent sensing systems" capable of recognizing and responding to changes in the chemical environment of the array in a dynamic manner. For example, spurious responses generated by interferents can be distinguished from true responses, and in some cases, it is possible to deconvolute contributions from primary ions and interferents in electrode signals and introduce automatic corrections to the analytical interpretation, leading to better accuracy in the results (see Figure 2.16).

Interest in electrode arrays has been stimulated by the availability of cheap, multichannel data acquisition cards for PCs, which is making this an increasingly attractive option compared to traditional single-electrode measurements with ion-meters.

2.10.1.   Practical Considerations

When employing sensor array–FIA, the ideal sensor characteristics can be quite different from those required when using a single electrode. To accurately estimate

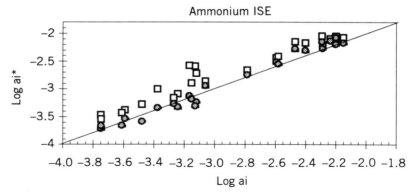

Figure 2.16. Improvement in predicted activities of NH_4^+ (log $ai*$) versus known activities (log ai) through use of an array approach. The corrected predictions (circles) are obviously nearer the ideal relationship (solid line) than the uncorrected (squares), which show a positive bias due to the widely varying background matrix, which cannot be detected and corrected for using a conventional single-electrode measurement method [42].

the selectivity coefficients, one must artificially use high interferent concentrations if the selectivity is good, in order to perturb the electrode signal sufficiently. Hence, the ideal sensor for these arrays is one that responds with moderate selectivity against the restricted number of ions implicitly included in the equations, and with extreme selectivity against all other interferents. The most important characteristic is stability, as the model characteristics should not vary significantly if the model is to be used in predictive mode to determine unknown ion concentrations. Some general points to note when using this approach are

- Calibration—a partial-factorial design is preferred using two levels for each ion for each electrode in the array. In general, this leads to l^{f-1} solutions for each electrode when l = number of levels (i.e., 2) and f = the number of factors (i.e., ions; 4 is a reasonable compromise). Hence for a 4-electrode array, 8 solutions are required to calibrate each electrode, and 32 solutions for the entire array. Clearly, the sensors should be very stable so that the relatively complex calibration does not have to be repeated very often (a one- or two-point check can be routinely used to monitor the calibration stability).

- FIA is convenient for this type of calibration as the large number of solutions can be quickly passed through the array and the calibration data collected. The model building procedure involves using an iterative nonlinear least-squares technique to obtain the set of model parameters that minimizes the cumulative error between the predicted and experimental potentials. Simplex optimization, projection pursuit [54–56] regression, and neural networks

[57,58] are examples of modeling methods that have been used for this purpose. Once these parameter estimations have been established, they can be used in a predictive mode to estimate the target ions in unknowns based on the potentials obtained from the electrode array.

Perhaps the most important advantage of the sensor array approach is that it is the only sensor-based method which can provide real-time multicomponent analytical information and simultaneously identify and compensate for matrix effects in a dynamic manner [59]. However, to make use of this advantage, it is crucial that the array model parameters remain stable during the period of use for obvious reasons.

A number of experimental approaches, including conventional batch methods [60], batch-injection analysis (BIA) [61], and FIA [59,62], have be used with ion-selective electrode arrays. However, given the number of solutions involved in calibrating even a modest 4-electrode array, FIA is to be preferred as it is the most amenable to automation. Table 2.3 compares the main characteristics of these approaches. The response of a FIA system with $4\times$ electrode array detector to duplicate injections a number of calibration solutions is illustrated in Figure 2.17. Even from a cursory visual examination, it is clear that the large response obtained with the Na, Li, and K electrodes in injections 1 and 2, and injections 5 and 6 suggests there is a high concentration of all three ions in this solution, and some degree of cross-response is to be expected. In contrast, the high response obtained with the K electrode in injections 2 and 3, and 7, and 8 suggests that this solution is predominately composed of potassium. This type of intelligent interpretation of the response can be achieved only through the use of an array approach.

Table 2.3. FIA, BIA, and Batch Process Main Characteristics[a]

Characteristics	Flow-Injection Analysis	Batch-Injection Analysis	Batch Processes
Reproducibility	✓✓✓	✓✓	✓
Steady-state measurements	✓✓	✓	✓✓✓
Kinetic measurements	✓✓	✓✓✓	✓
Analysis time	✓✓	✓✓✓	✓
Sample throughput	✓✓✓	✓✓	✓
Reagent consumption	✓	✓✓	✓✓✓
Automation	✓✓✓	✓✓	✓

[a]The techniques best suited for sensor array analysis clearly are flow-injection, for fundamental and kinetic studies; and batch injection, mainly for kinetic measurements. FIA is the only one that can be automated completely using reasonable resources.

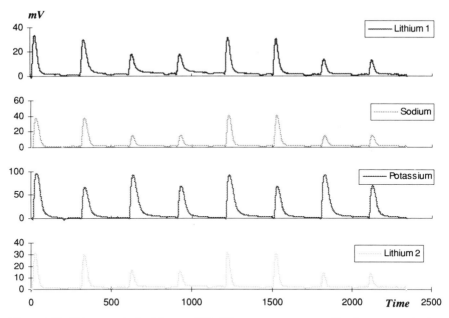

Figure 2.17. FIA traces obtained from an ISE–FIA–sensor array system. Simultaneous monitoring each ISE permits multivariate analysis and interferent compensation to be implemented on-line. Each ISE trace shows the different "high" and "low" responses to different levels of ions Li^+, Na^+, and K^+ in ternary solutions designed to calibrate a sensor array optimized for analysis of lithium in blood samples. Comparison of the relative heights of equivalent peaks in each trace enables cross-responses to be identified.

2.10.2.   Self-Diagnosing Systems

Full computer control and automation of the FIA–sensor array system opens the possibility of building "intelligent" subroutines within the control procedures that can enable the detection of system malfunction or sensor degradation.

This approach may require a number of identical sensors with the same response characteristics. Comparison of the response pattern and prediction polling characteristics, both between the equivalent ISEs and among all the sensor array sensors, can identify any malfunctioning sensor, and its response can be automatically discarded and a warning flagged.

Other problems arising during the operation of the FIA system can be diagnosed, including

- A drop in the flow rate because of a leakage in the tubing can be detected by an abnormal increase in the analysis time in all the sensors.

- Total lack of signal from the sensor array will indicate that the system is failing to properly inject the sample into the carrier stream (except for the unlikely case that the sample has exactly the same composition as the carrier).
- Large artifacts occurring only at the predicted peak location indicate that the electrochemical cell is electronically open, probably because of the introduction of air during injection.
- If the above happens continuously in one detector channel it suggests that an air bubble is trapped at that sensor location.

If the sensor array contains redundant ISEs (i.e., multiple electrodes for the same primary ion), diagnosis on the working characteristics of the electrode membranes can be obtained by comparing the responses, for example

- A reduced signal from the sensor will indicate that the particular sensor is past its useful working life and has to be replaced; that the complete sample volume does not reach the sensor site, perhaps caused by leakage; or that there is a faulty electrical connection in that particular electrode.
- Also, selective detector poisoning is suggested when only one type of the redundant sensors continuously fails to give a signal.

Together, when combined into an overall system, the user will have an instrument capable of real-time self-diagnosis and analytical interpretation, which will lead to more user confidence in the results. The real challenge is for researchers is to produce compact, solid-state, highly stable chemical sensors with reproducible characteristics (cell constant, slope, and selectivity). The application of planar fabrication technologies used for thin- and thick-film deposition (e.g., screen printing, chemical vapor deposition) to electrode construction will make an important contribution to meeting this challenge (see Section 2.11).

2.11. SOLID-STATE ISEs

There has been considerable interest in the application of planar fabrication techniques to the manufacturing of ISEs. During the 1980s, much effort was expended toward the development of silicon-based devices known as *ion-selective field-effect transistors* (ISFETs), which attempted to integrate ISE-PVC membranes with FET technology. However, the results were largely disappointing because of the problems in establishing a stable internal reference potential where the PVC membrane came in contact with the FET surface. While activity in this area has greatly diminished, research is still ongoing, and perhaps the advent of better fabrication technologies, surface analysis, and visualization techniques (e.g., atomic-force microscopy) and a growing understanding of the fundamental materials problems involved may lead to renewed interest [63].

More recently, much attention has focused on the use of materials that may be able to provide the mixed electronic–ionic conductance required for purpose. The problem is that the electrode-filling solution simultaneously provides a stable internal membrane boundary potential and a stable internal reference electrode potential, but the use of the filling solution is incompatible with planar fabrication techniques such as screen printing. Placing the ion-sensitive membrane directly onto a metallic or Ag/AgCl surface leads to a blocked interface to charge transfer, which gives rise to severe drift, and irreproducible, noisy signals (see Figure 2.18).

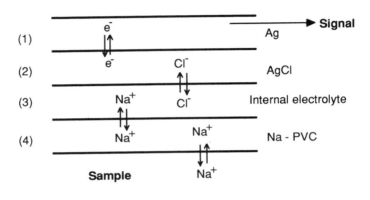

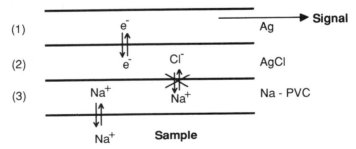

Figure 2.18. (*Top*) in a conventional liquid-filled electrode, the PVC exchanges ions with the sample solution (Na^+-selective in this example), and the internal electrolyte serves a dual role by (i) stabilizing the internal boundary potential of the PVC membrane by exchanging charge (Na^+ ions) and (ii) stabilizing the internal reference electrode potential by exchanging Cl^- ions with the AgCl layer. The AgCl converts this to electronic charge transfer via exchange of electrons (e^-) with the Ag film, and the driving force behind this overall process is picked up as a potential by the external circuit. (*Bottom*) If the ion-sensitive PVC membrane is deposited directly on a metal or Ag/AgCl substrate, a blocked interface is generated at the substrate–membrane boundary due to the lack of a common mechanism for charge transfer (see Fig. 2.20).

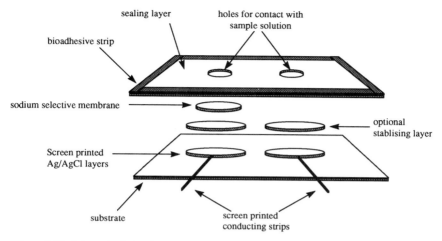

Figure 2.19. Schematic design for a planar/solid-state combination ion-sensitive electrode (left) and reference electrode (right) for applications involving surface ion measurements. The sensing electrode is composed of screen-printed Ag and AgCl films, an "unblocking" film to enable efficient exchange of charge between the PVC membrane and the AgCl surface, and an ion-sensitive PVC membrane. The reference electrode consists of screen-printed Ag and AgCl films. If the AgCl film is left in open contact with the sample, the chloride content must be constant unless the reference half-cell is protected (stabilized) with a further film (e.g., KCl-doped resin) that can simultaneously exchange Cl^- with the AgCl substrate and provide a stable contact potential with the sample.

Materials like epoxies [64], conducting polymers [65], and salt-doped resins [66] have been examined as possible solid-state substitutes for the internal filling solution in electrodes. Hydrogels have also been shown to be useful solid-like materials for entrapping salts such as KCl between the sensing membrane and the internal reference electrode surface. Many hydrogels are now available, some of which (through crosslinking) do not exhibit the swelling that caused problems in earlier designs. The recent paper by Crosfet and co-workers demonstrated the utility of fabricating small, solid-state arrays of ISEs using salt-doped poly-hema as a 'solid-like' film between an ion-sensitive PVC film and the Ag/AgCl substrate. These arrays were used to monitor K^+ fluxes on the surface of a beating heart [67]. A schematic design for a screen-printed planar ISE-reference electrode cell is illustrated in Figure 2.19. A hydrogel is used to entrap a salt (in this case, NaCl) between the AgCl internal planar reference electrode and a Na-selective PVC film. The external reference electrode is a second planar screen-printed Ag/AgCl electrode that can be left bare for applications such as the determination of NaCl in sweat, or coated with a salt-doped polymer or resin as mentioned above to produce a Na-sensitive cell. Figure 2.20 shows some results obtained with this type of cell. A stable signal is evident with a fast response and approximately theoretical slope.

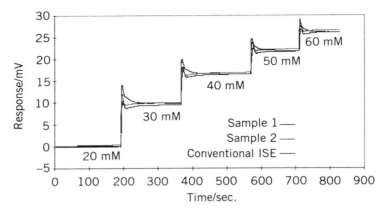

Figure 2.20. Comparison of the performance of two screen-printed sodium-selective sensors (samples 1 and 2) and an equivalent liquid-filled electrode (conventional ISE). The responses are essentially the same, showing that the NaCl-doped hydrogel intermediate layer is satisfactory substitute for the internal filling solution (concentrations are for final concentrations of Na^+ after spiking).

Further refinement of the materials and design should lead to very cheap ion sensors with reproducible characteristics.

For further background on solid-state sensors, the reader is referred to the excellent text by Madou and Morrison [68].

2.12. ION-SELECTIVE OPTODES

The receptors used in potentiometric sensors can be induced to generate a visible color change on complexation with metal ions by means of acidochromic dyes. Desirable properties for these dyes can be summarised as follows:

- Absorb strongly in the visible region
- Possess weak acid nature
- Absorbance spectrum greatly affected by protonation–deprotonation
- Lipophilic in character to reduce leaching from lipophilic membranes to a minimum
- Chemically and photochemically stable

2.12.1. Mechanism of Optical Response

As outlined in Section 2.4.5, the performance of cation-selective PVC membrane potentiometric sensors is determined by the physicochemical properties of the ligand/receptor immobilized in the membrane. Normally, these function by complexation of metal ions in the boundary region where the essentially lipophilic PVC–plasticizer phase merges with the aquosample phase. Hence metal ions are drawn into the PVC–plasticiser phase (m), and, in order to preserve charge neutrality, two mechanisms may occur:

1. Anions $[A^-_{(aq)}]$ are coextracted from the sample phase:

$$M^+_{(aq)} + A^-_{(aq)} + L_{(m)} \rightleftharpoons LM^+_{(m)} + A^-_{(m)} \qquad (2.44)$$

2. Cations (e.g., H^+) are expelled from the membrane phase:

$$M^+_{(aq)} + CromH^+_{(m)} + L_{(m)} \rightleftharpoons LM^+_{(m)} + Chrom_{(m)} + H^+_{(aq)} \qquad (2.45)$$

In the case of ISEs, mechanism 1 often applies, but for optodes, mechanism 2 is required, as this couples deprotonation of the acidic form of the chromophore in the membrane phase $[CromH^+_{(m)}]$ and subsequent release of protons to the sample phase with the complexation of the metal ion by the ligand. Clearly, while this results in the optical transduction of the complexation reaction, it also introduces an undesirable pH-dependent process into the overall response scheme. Therefore, the sample phase should be carefully buffered in order to prevent membrane colour changes arising from fluctuations in the sample pH, and also to provide a "sink" for the protons released from the membrane. In consequence, such systems are in general not suited continuous monitoring situations where the sensor is immersed in a sample whose composition may be dynamically changing. However, they are ideal for one-shot measurements using simple or no instrumentation (i.e., visual indication [69]) such as screening analysis performed in the field or at remote locations. The utility of this approach was very elegantly demonstrated by Seitz [70] and Wolfbeis [71], and the principles of the measurements outlined by Simon [72] and co-workers. A suitable dye material, ETH5294 (Fluka), is illustrated in Figure 2.21. The fact that the dye is positively charged in the acidic form means that a lipophilic counterion is required to be present in the membrane phase for charge neutrality. The lipophilic counterions conventionally used in ISE membranes (tetraphenylborate or tetra-p-chlorophenylborate) are unsuitable for this purpose as they rapidly decompose during use because of the presence of photons and protons in the membrane. This problem can be overcome by the use of more photostable lipophilic anions such as the Kobayashi reagent [NaTm(CF3)phenyl borate] [73].

2.12.2. Positively Charged Acidochromic Dyes

The UV–vis spectra of the acidic and basic forms of the dye are illustrated in Figure 2.22. Incorporation of this dye into a plasticized PVC membrane with the anionic counterion and a calix[4]arene tetraphosphine oxide calcium-selective ligand

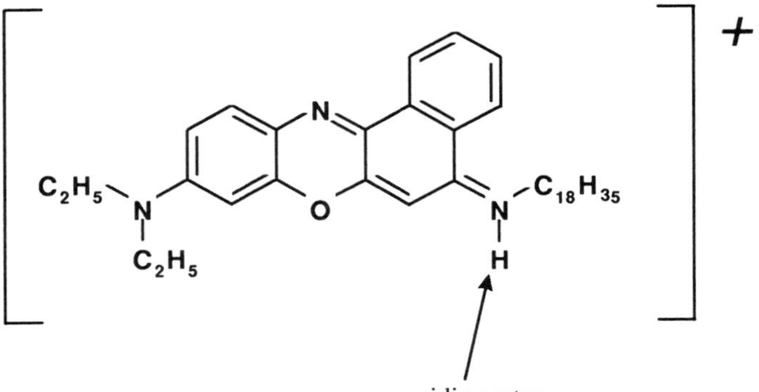

Figure 2.21. Structure of the lipophilic analog of the pH indicator Nile Blue (ETH5294).

(see Figure 2.23) leads to a PVC membrane that can reproduce the selectivity of the equivalent potentiometric PVC membrane based on the same ligand. The potentiometric selectivity for calcium ions against sodium ions is convincingly illustrated in Figure 2.24. In a similar manner, Figure 2.25 shows that the UV–vis spectrum changes markedly as calcium ions are complexed by the ligand and protons are expelled from the membrane to preserve charge neutrality whereas a much reduced optical response is obtained when the same membrane is exposed to equivalent concentrations of sodium ions. The great advantage of this approach is that a range of optical equivalents to the well-known PVC membrane ISEs can be

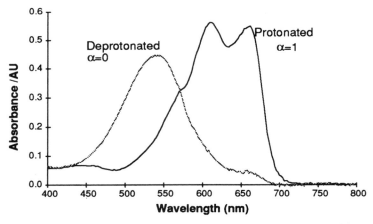

Figure 2.22. UV–vis Spectra of ETH5294 in the protonated and deprotonated forms.

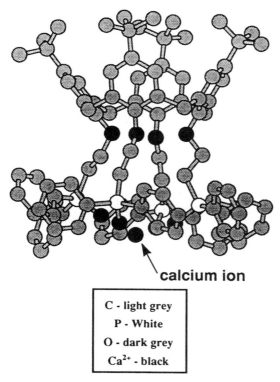

calcium ion

C - light grey
P - White
O - dark grey
Ca^{2+} - black

Figure 2.23. Energy-minimized structure of tetraphosphineoxide calix[4]arene: calcium complex calculated using Hyperchem v. 4.0. This ligand shows good selectivity for calcium over sodium ions.

rapidly developed by substituting the various ligands used in successful potentiometric sensors.

2.12.3. Neutral Acidochromic Dyes

An alternative strategy is to use a dye that is neutral in the acidic form (ChromH). Unlike the previous case, this type of indicator does not need a counter ion present in the membrane. In this case, the mechanism is

$$M^+_{(aq)} + ChromH_{(m)} + L_{(m)} \rightleftharpoons LM^+_{(m)} + Chrom^-_{(m)} + H^+_{(aq)} \qquad (2.46)$$

In addition, the indicator (ChromH) can be covalently bound to the ligand (L), and therefore only a single membrane component is required. The structure of one such

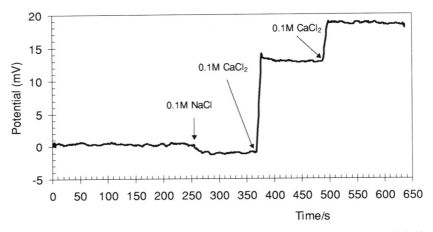

Figure 2.24. Transient responses of an ISE based on tetraphosphineoxide calix[4]arene to an injection of 250 μL of NaCl followed by two further injections of 250-μL aliquots of 0.1 M CaCl$_2$. A much larger response to CaCl$_2$ is evident, reflecting the selectivity of the ligand.

ligand is illustrated in Figure 2.26. It consists of a calix[4]arene tetraester macrocycle similar to that shown in Figure 2.23, with nitrophenylazophenyl acidochromic indicator moeties attached near the ion-complexing cavity defined by the carbonyl and phenyl oxygen atoms. Figure 2.27 shows the response obtained with a solution of this ligand in chloroform to spikes of LiClO$_3$ in the presence of tri-n-didodecylamine, which acts as a weak base and picks up the free protons released on complexation of the lithium ion by the ligand. As this is a single-phase measurement, the base must be added. This base should not be strong enough to deprotonate the free ligand to any great extent. When Li$^+$ ions are added, they are complexed by the ligand, and the presence of the positively charged ion near the chromophore has the effect of increasing its acidity, and proton transfer to the base occurs.

This effect can be used in reverse to detect volatile bases (B) such as ammonia and simple amines. The overall reaction in this case is

$$L - CromH_{(m)} + B_{(g)} \rightleftharpoons L - Chrom^-_{(m)} + BH^+_{(m)} \qquad (2.47)$$

However, it can be predicted that the sensitivity of this response can be enhanced by using the complex (LM$^+$ − ChromH) rather than the free ligand (L − ChromH) as the indicator;

$$LM^+ - ChromH_{(m)} + B_{(g)} \rightleftharpoons LM^+ - Chrom^-_{(m)} + BH^+_{(m)} \qquad (2.48)$$

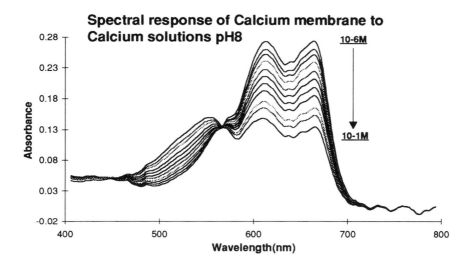

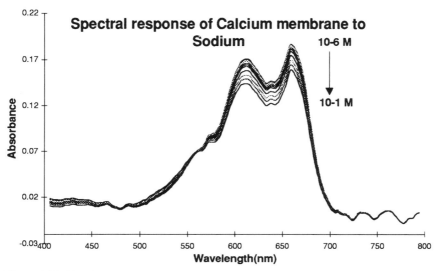

Figure 2.25. UV–vis spectra of PVC membranes containing the ligand tetraphosphineoxide ca-lix[4]arene, ETH5294, and an ion exchanger with the aqueous phase at pH 8 on exposure to 10^{-6}–0.1 M Ca^{2+} (*top*) and Na^+ (*bottom*). Clearly the response is much larger for Ca^{2+} than Na^+ ions, demonstrating that the potentiometric selectivity is reproduced in the optical response.

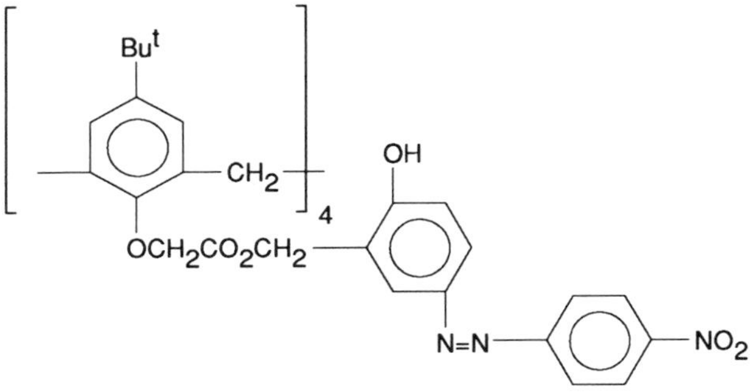

Figure 2.26. Structure of the tetraazophenolnitrophenol calix[4]arene chromionophore.

This effect is nicely illustrated in Figure 2.28, which shows the response of a fiber-optic probe coated with PVC films containing the free ligand (*a*) and the Li complex (*b*) of the calix[4]arene nitrophenylazophenyl chromionophore [74]. Clearly the response is much more sensitive in the case of the latter because of the increased acidity of the complex. Other results suggest that the effect can be tuned by varying the proportion of complex to free ligand, or by varying the ion used to form the complex. This type of behavior raises the possibility of generating mate-

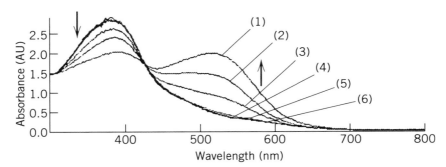

Figure 2.27. Changes in the UV–vis spectrum of tetraazophenolnitrophenol calix[4]arene tetraacetate to spikes of LiClO$_3$. The reaction is carried out in THF. The experiment was carried out by adding incremental spikes of aqueous LiClO$_4$ to 2.5 mL of a 5 × 10^{-5} M solution of the nitrophenylazophenyl calix[4]arene chromionophore in THF in the presence of 100 μL of TDDA to give the following final Li$^+$ concentrations: *A*, 0.1 M; *B*, 0.01 M; *C*, 10^{-3} M; *D*, 10^{-4} M; *E*, 10^{-5} M; *F*, 10^{-6} M; *G*, 0.0 (blank) M.

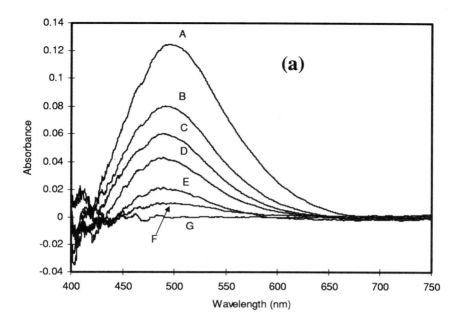

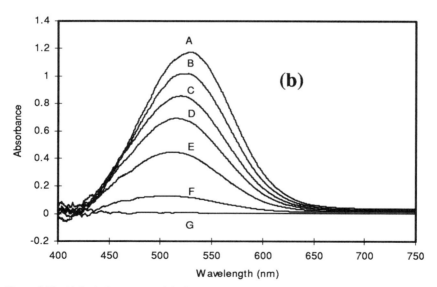

Figure 2.28. (*a*) Optical responses of the free tetraazophenolnitrophenol calix[4]arene tetraacetate ligand to gaseous ammonia: *A*, 500 ppm; *B*, 200 ppm; *C*, 100 ppm; *D*, 50 ppm; *E*, 20 ppm; *F*, 5 ppm; *G*, N$_2$. (*b*) Optical responses of the lithium complex to gaseous ammonia: *A*, 100 ppm; *B*, 50 ppm; *C*, 20 ppm; *D*, 10 ppm; *E*, 5 ppm; *F*, 2 ppm; *G*, N$_2$. A scan of the coated fibre in nitrogen is used as a reference. [74]

rials with customized optical linear range and sensitivity, depending on the concentration span and/or limit of detection and sensitivity required by an application. Similar responses can be obtained from other volatile bases such as aliphatic amines [75].

2.13. CONCLUDING REMARKS

The development of new methods for ion sensing continues to be an area of interest research interest. We can expect progress in the use of novel automated fabrication techniques, planar sensor designs, sensor arrays incorporating anion as well as cation sensors, novel receptors for anions, and optically responsive equivalents of the well-known potentiometric devices, all of which will greatly broaden the application base for these sensors.

REFERENCES

1. Wipf, D. O., and Bard, A. J., *Analyst* **119,** 719–726 (1994).
2. Wipf, D. O., and Bard, A. J., *J. Electrochem. Soc.* **138,** 469–474 (1991).
3. Pranitis, D. M., Telting-Diaz, M., and Meyerhoff, M. E., *Crit. Rev. Anal. Chem.* **23,** 163–186 (1992).
4. Janata, J., Josowicz, M., and DeVaney, D. M., *Anal. Chem.* **66,** 207R–228R (1994).
5. Janata, J., *Anal. Chem.* **64,** 196R–219R (1992).
6. Janata, J., *Anal. Chem.* **62,** 33R–44R (1990).
7. Solsky, R. L., *Anal. Chem.* **62,** 21R–33R (1990).
8. Wang, J., *Anal. Chem.* **65,** 450R–453R (1993).
9. Crow, D. R., *Principles and Applications of Electrochemistry,* 4th ed., Blackie Academic & Professional, Glasgow, UK, 1994.
10. Morf, W. E., *The Principles of Ion-Selective Electrodes and of Membrane Transport,* Elsevier Scientific, Amsterdam, 1981.
11. Suzuki, K., Aruga, H., and Shirai, T., *Anal. Chem.* **55,** 2011–2013 (1983).
12. Yim, H. S., Cha, G. S., and Meyerhoff, M. E., *Anal. Chim. Acta* **237,** 115–125 (1990).
13. Koryta, J., Dvorák, J., and Kavan, L., *Principles of Electrochemistry,* 2nd ed., Wiley, Chichester, UK, 1993, Chapter 6.
14. Vogel, A. I., *Textbook of Quantitative Chemical Analysis,* 5th ed., Longman Scientific & Technical, Essex, England, 1989, Chapter 15, pp. 548–590.
15. Pranitis, D. M., and Meyerhoff, M. E., *Anal. Chem.* **59,** 2345–2350 (1987).
16. Fraticelli, Y. M., and Meyerhoff, M. E., *Anal. Chem.* **53,** 992–997 (1981).
17. Beer, P. D., *Chem. Commun.* **1996,** 689.
18. Park, S. B., Matuszewski, W., Meyerhoff, M. E., Liu, Y. H., and Kadish, K. M., *Electroanalysis* **3,** 909 (1991).

19. Wuthier, U., Hung-Viet-Pham, Zeund, R., Welti, D., Funck, R. J. J., Bezegh, A., Ammann, D., Pretsch, E., and Simon, W., *Anal. Chem.* **56,** 535 (1984).

20. Badr, I. H. A., Meyerhoff, M. E., and Hassan, S. S. M., *Anal. Chem.* **67,** 2613–2618 (1995).

21. Badr, I. H. A., Meyerhoff, M. E., and Hassan, S. S. M., *Anal. Chim. Acta* **310,** 211–221 (1995).

22. Allen, J. R., Cynkowski, T., Desay, J., and Bachas, L. G., *Electroanalysis* **4,** 533–537 (1992).

23. Forster, R. J., Cadogan, A., Telting, Diaz, M., Diamond, D., Harris, S., and McKervey, M. A., *Sensors and Actuators B* **4,** 325–331 (1991).

24. Cunningham, K., Svehla, G., Harris, S. J., and McKervey, M. A., *Analyst* **118,** 341–345 (1993).

25. Cadogan, A. M., Diamond, D., Smyth, M. R., Deasy, M., McKervey, M. A., and Harris, S. J., *Analyst* **114,** 1551–1554 (1989).

26. Ozawa, S., Hauser, P. C., Seile, K., Tan, S. S. S., Morf, W. E., and Simon, W., *Anal. Chem.* **63,** 640–644 (1991).

27. Pranitis, D. M., and Meyerhoff, M. E., *Anal. Chem.* **59,** 2345–2350 (1987).

28. Fraticelli, Y. M., and Meyerhoff, M. E., *Anal. Chem.* **53,** 992–997 (1981).

29. Davies, O. G., Moody, G. J., and Thomas, J. D. R., *Analyst* **113,** 497–500 (1988).

30. Forster, R. J., and Diamond, D., *Anal. Chem.* **64,** 1721–1728 (1992).

31. Otto, M., and Thomas, J. D. R., *Ion-Selective Electrode Rev.* **8,** 55–84 (1986).

32. Ramette, R. W., *Chemical Equilibrium and Analysis,* Addison-Wesley, Reading, MA, 1981, p. 95.

33. Christian, G. D., *Analytical Chemistry,* 4th ed., Wiley, New York, 1986, p. 110.

34. Davey, D. E., Mulcahy, D. E., and O'Connell, G. R., *Electroanalysis* **5,** 581–588 (1993).

35. Meyerhoff, M. E., and Kovach, P. M., *J. Chem. Educ.* **60,** 766–768 (1983).

36. van Standen, J. F., *Anal. Proc.* **24,** 331–333 (1987).

37. Douglas, J. G., *Anal. Chem.* **61,** 922–924 (1989).

38. Hungerford, J. M., and Christian, G. D., *Anal. Chim. Acta* **200,** 1–19 (1987).

39. Hungerford, J. M., Christian, G. D., Ruzicka, J., and Giddings, J. C., *Anal. Chem.* **57,** 1794–1798 (1985).

40. Whitman, D. A., Seaholtz, M. B., Christian, G. D., Ruzicka, J., and Kowalski, B. R., *Anal. Chem.* **63,** 775–781 (1991).

41. Frenzel, W., *Analyst* **113,** 1039–1046 (1988).

42. Sáez de Viteri, F. J., and Diamond, D., *Analyst* **119,** 749–758 (1994).

43. Ruzicka, J., *Anal. Chem.* **55,** 1040A–1053A (1983).

44. Ranger, C. B., *Anal. Chem.* **53,** 20A.

45. Ruzicka, J., and Hansen, E. H., *Anal. Chim. Acta* **114,** 19–44 (1980).

46. Diamond, D., and Forster, R. J., *Anal. Chim. Acta* **276,** 75–86 (1993).

47. Lewenstam, A., Saarinen, K. S., and Hulanicki, A., *IFCC Workshop* 1986.

48. MacCarthy, P., Klusman, R. D., and Cowling, S. W., *Anal. Chem.* **63,** 301R–342R (1991).

49. Fuckskó, J., Tóth, K., and Pungor, E., *Anal. Chim. Acta* **194,** 163–170 (1987).

50. Pungor, E., Tóth, K., and Hrabéczy-Páll, A., *Trends in Anal. Chem.* **3,** 28–30 (1984).

51. Suzuki, K., Aruga, H., and Shirai, T., *Anal. Chem.* **55,** 2011–2013 (1983).

52. Nagels, L. J., and De Backer, B. L., *Anal. Chem.* **68,** 4441 (1996).

53. Hauser, P. C., Renner, N. D., and Hong, A. P. C., *Anal. Chim. Acta.* **295,** 181 (1994).

54. Friedman, J. H., and Stuetzle, W., *J. Am. Stat. Assoc.* **76,** 817–823 (1981).

55. Beebe, K. R., and Kowalski, B. R., *Anal. Chem.* **60,** 2273–2278 (1988).

56. Sáez de Viteri, F. J., and Diamond, D., *Electroanalysis* **6,** 9–16 (1994).

57. Hartnett, M., Diamond, D., and Barker, P. G., *Analyst* **118,** 347–354 (1993).

58. Janson, P. A., *Anal. Chem.* **63,** 357A–362A (1991).

59. Diamond, D., and Forster, R. J., *Anal. Chim. Acta* **276,** 75–86 (1993).

60. Forster, R. J., Reagan, F., and Diamond, D., *Anal. Chem.* **63,** 876–882 (1991).

61. Lu, J., Chen, Q., Diamond, D., and Wang, J., *Analyst* **118,** 1131–1135 (1993).

62. Forster, R. J., and Diamond, D., *Anal. Chem.* **64,** 1721–1728 (1992).

63. Hogg G., Lutze, O., and Cammann, K., *Anal. Chim. Acta* **335,** 103 (1996).

64. Yun, S. Y., Hong, Y. K., Oh, B. K., Cha, G. S., Nam, H., Lee, S. B., and Jin, J.-I., *Anal. Chem.* **69,** 868 (1997).

65. Cadogan, A., Gao, Z., Lewenstam, A., Ivaska, A., and Diamond, D., *Anal. Chem.* **64,** 2496 (1992).

66. Rehm, D., McEnroe, E., and Diamond, D., *Anal. Proc. Comm.* **32,** 319 (1995).

67. Crosfet, V. V., Erdosy, M., Johnson, T. A., Buck, R. P., Ash, R. B., and Neumann, M. R., *Anal. Chem.* **67,** 1647 (1995).

68. Madou, M. J., and Morrison, S. R., *Chemical Sensing with Solid State Devices,* Academic Press, Boston, 1989.

69. Edmonds, T. E., Lee, J. M., and Lee, J. D., *Anal. Commun.* **1997,** 1H.

70. Zhang, Z., Mulin, J. L., and Seitz, W. R., *Anal. Chim. Acta* **184,** 251 (1986).

71. Schaffar, B. H., and Wolfbeis, O. S., *Anal. Chim. Acta* **217,** 1 (1989).

72. Seiler, K., and Simon, W., *Anal. Chim. Acta* **266,** 73 (1992), and references cited therein.

73. Seiler, K., *Ion-selective Optode Membranes,* Fluka Chemie AG, 1993, pp. 38–39.

74. Grady, T., Butler, T., MacCraith, B., McKervey, M. A., and Diamond, D., *Analyst* **122,** 803–806 (1997).

75. McCarrick M., Harris S. J., and Diamond D., *J. Mater. Chem.* **4,** 217–221 (1994).

CHAPTER

3

AMPEROMETRIC METHODS OF DETECTION

JOHN F. CASSIDY

Department of Chemistry, Dublin Institute of Technology, Dublin, Ireland

ANDREW P. DOHERTY[1] and JOHANNES G. VOS

School of Chemical Sciences, Dublin City University, Dublin, Ireland

3.1. INTRODUCTION

An amperometric sensor is based on the principle that when a species is oxidized or reduced at an electrode, the current produced is directly related to the concentration of the species. Amperometric sensors represent a subsection of electrochemistry that has long been applied to analytical chemistry. With the developments in potentiometry and polarography, there is a wide range of instrumental applications of these techniques that can be applied to analysis. From the outset, Faradaic electrochemistry was plagued with problems of electrode stability as many electrochemical processes caused passivation of the electrode surface. However with the advent of the dropping-mercury electrode (DME), which provided a renewable surface, a revolution occurred in analytical electrochemistry. Unfortunately, the inherent difficulties with mercury electrodes (toxicity, liquid, and limited anodic potential range), along with the advances made in selective spectroscopic techniques (atomic absorption spectroscopy), caused a decline in the interest in Faradaic electrochemistry as an analytical technique. Recently, however, with the development of new pulsed measurement techniques coupled with more user-friendly computer-driven equipment, a revival in electroanalysis has started. Other important developments were the first enzyme electrode in the late 1960s and the introduction of chemically modified electrodes in the early 1970s. These developments were aimed at improving the selectivity of analytical techniques by designing analyte-specific surfaces. With this concept, tailoring the properties of conventional electrode surfaces has entered the arena and further helped the renaissance of electroanalytical chemistry. The prospect of designer electrodes is

[1]Present address: Chemistry Department, Bedson Building, University of Newcastle, Newcastle upon Tyne, NE1 7RU, UK.

becoming a reality and the possibility of commercial returns is creating the impetus for modified electrode research.

This chapter deals with the fundamental theory of the electrochemistry underlying amperometric sensors, detailing some of the more popular instrumental methods. Attention will be paid to the different measurement techniques that have been developed to increase the sensitivity of the measurement. Finally the basic principles of modified electrodes, including biosensors, will be addressed. Examples of the analytical application of the different measurement techniques and some modified electrodes will be given.

Modified electrodes can be crudely divided into two groups: biologically modified electrodes and chemically modified electrodes (CME). A chemically modified electrode has a purely synthetically controlled interface, whereas a biologically modified electrode contains a biologically active component for analyte recognition. Here amperometric sensors will be discussed, along these two divisions and further subdivisions made where necessary.

3.1.1. Electrochemical Fundamentals

Electrochemistry is primarily an interfacial process involving the transfer (or impending transfer) of an electron from a species in solution to an electrode (or vice versa). In order to use electrochemistry as an analytical tool, there has to be a contact between the electrode and the analyte for the electrode to sense the species. In order to understand the two main applications of electrochemistry for analysis—potentiometry and amperometry—it is necessary to briefly discuss what happens when an electrode is placed in an electrolyte solution.

Before tackling this concept, the term potential (E) has to be defined. The potential of a solution containing the reduced form of a couple Red and the oxidized form Ox, is easiest understood in terms of the Nernst equation as follows:

$$E = E^{0'} + \frac{RT}{(nF)} \ln\left\{\frac{[Ox]}{[Red]}\right\} \tag{3.1}$$

where R is the gas constant, ($8.314 \text{ J mol}^{-1} \text{ K}^{-1}$), T is the temperature in Kelvin, F is the charge corresponding to one mole of electrons ($96487 \ C$), and [Ox] and [Red] represent the concentration of the species in moles per cubic decimeter. The formal potential ($E^{0'}$) is a parameter characteristic of a couple:

$$Ox + ne^- = Red \qquad E^{0'} \tag{3.2}$$

A widely studied couple is the ferrocene (fc)/ferricinium (fc$^+$) couple, where there is a single electron transfer involved:

$$\left[\begin{array}{c}\bigcirc\\Fe\\\bigcirc\end{array}\right]^{+} + e \rightleftharpoons \begin{array}{c}\bigcirc\\Fe\\\bigcirc\end{array} \quad E_1^{o'} \tag{3.3}$$

This system, along with a series of derivatives, is popular because both the reduced form (ferrocene) and the oxidized form (ferricinium) are both reasonably stable in solution, which means that they can be readily interconverted.

If a solution is prepared consisting of 0.01 mol/dm^3 ferrocene (fc) and 0.01 mol/dm^3 ferricinium ion (fc$^+$) in acetonitrile, the potential of this solution according to the Nernst equation would be

$$E = E_1^{0'} + \frac{RT}{(nF)} \ln\left\{\frac{[\text{fc}^+]}{[\text{fc}]}\right\} = E_1^{0'} \tag{3.4}$$

Thus the formal potential of a couple may be defined as the potential of a solution containing an equal concentration of oxidized and reduced forms of the couple. For other ratios of concentrations, the solution potential may be calculated using the Nernst equation.

3.1.2. Potential

If an inert electrode, such as platinum or gold, were placed in such a solution, the electrode immediately adopts the potential of that solution. If the electrode were kept at some other potential, work would have to be done in to maintain this potential difference. Potential is not usually measured as an absolute magnitude but in relative terms, and so it is normal to employ benchmarks, (which themselves have a constant potential) with respect to which potentials are quoted. These benchmarks or reference electrodes consist of an electrode system comprising oxidized and reduced forms of a couple. The originally widely used reference electrode consisted of a platinum electrode placed in a 1 mol/dm^3 H$^+$ solution that was bubbled with H$_2$ gas [standard hydrogen electrode (SHE)]. This was found to be impractical for frequent use. Presently used reference electrodes are generally chosen to be robust and may consist of Hg/Hg$_2$Cl$_2$/Cl$^-$ or Ag/AgCl/Cl$^-$ couples as shown in Figure 3.1a. Thus the potential of the ferrocene/ferricinium solution may be measured as shown in Figure 3.1b.

This can be seen as a complete circuit where the potential difference between the solution (which is the same as that of the platinum or working electrode), and the reference electrode may be measured. An important consequence of this is

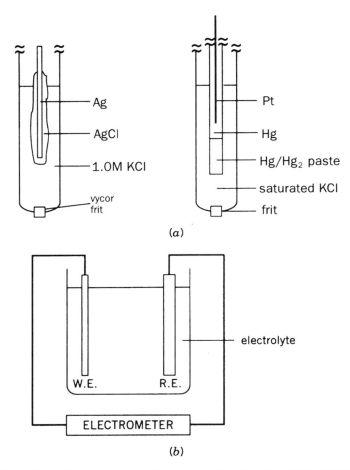

Figure 3.1. (*a*) Sketch of a silver/silver chloride electrode ($E = 222$ mV vs. SHE (1 M KCl), $E = 197$ mV vs. SHE, saturated KCl) and a calomel electrode ($E = 242$ mV vs. SHE (saturated KCl)). (*b*) A simple cell for measuring solution potential. The electrolyte consists of a solution of a dissociated salt that reduces the resistance of the system.

that the reported value of the potential (usually in volts) is meaningless unless the type of reference electrode used is also stated. It can be seen that according to the Nernst equation the potential of the inert electrode [termed the *working electrode* (WE)] in Figure 3.1*b* and the natural log of the ferrocene concentration are linearly related.

For the Nernst equation to be valid, a prerequisite is that no current passes through the circuit. In practice, the current that passes is very small. If there were a passage of current, then either ferrocene could be oxidized or the ferricinium reduced, in which case their concentrations would change.

This analytical technique of potentiometry (relating potential to analyte concentration) suffers from the problem of specificity. The potential of a solution with a mixture of two couples would be averaged out since it depends on both couples. A degree of selectivity is obtained by introducing an ion-selective membrane that is positioned between two reference electrodes. In this manner various charged species may be analysed using a modified form of the Nernst equation. The most popular and effective example of a wide range of ion-selective membranes is the glass membrane, which shows a specificity for the hydrogen ion. Rather than using a glass membrane, a wide variety of solid-state and liquid membranes are employed in potentiometry, details of which are to be found in the chapter on potentiometry.

3.2. THEORY AND PRACTICE OF AMPEROMETRY

3.2.1. Theory

Rather than obtaining analytical information from a system at equilibrium (i.e., no current passed) as described in potentiometry, one can also shift the equilibrium and relate the concentration of analyte to the number of electrons transferred across an electrode solution interface that is observed as a current. Let us, for example, consider the ferrocene/ferricinium system in Eq. (3.3). It is possible to shift this reaction to the left by applying a potential which is more positive than $E_1^{0'}$ to the platinum electrode in Figure 3.1b. The resulting current is then related to the fc concentration. This approach introduces considerable difficulties since the analyte is consumed at the electrode as can be seen in Figure 3.2.

When such a potential is applied to the platinum electrode, any fc close to the electrode will be oxidized (see Figure 3.2). As time progresses, the current will decay gradually as all the fc close to the electrode is used up. Thus, at first sight, this method does not appear to be useful from an analytical point of view since the current decreases with time.

In order to relate current to concentration in solution, it is convenient to consider two limiting cases. When ferrocene is oxidized at the electrode, the current obtained may be limited by either or both of two processes, namely, the kinetics of electron transfer at the electrode or mass transport of ferrocene toward the electrode surface.

It is possible from thermodynamic considerations to develop an equation for current under conditions where the kinetics are the rate-determining step. This cur-

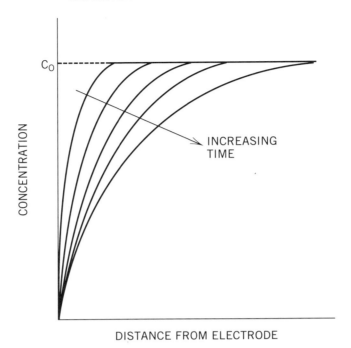

DISTANCE FROM ELECTRODE

Figure 3.2. Representation of concentration–distance profiles when a potential step is applied so that the concentration of ferrocene at the electrode surface falls to zero. This leads to a difference in concentration and the ferrocene diffuses in from bulk solution to remove this concentration difference. Over a period of time this process can be seen as an expanding depletion (or diffusion) layer.

rent is traditionally given as the net current difference between cathodic (reduction) and anodic (oxidation) processes: $i = i_{\text{cathodic}} - i_{\text{anodic}}$. This may be expanded to give

$$i = nFAk^0\{\exp(-\alpha nF(E - E^{0\prime})/RT)C_0(0,t)$$
$$-\exp((1 - \alpha)nF(E - E^{0\prime})/RT)C_r(0,t)\} \qquad (3.5)$$

where k^0 is the standard heterogeneous rate constant for the redox process at the electrode surface; α is the transfer coefficient, which is a measure of how much the applied potential is helping the forward reaction and hindering the backward reaction; α is usually taken as 0.5 for an ideal reaction; n is the number of electrons transferred; A is the electrode area; and F is the number of coulombs per mole of electrons. The concentrations are specified as those at the electrode surface where $C(x,t)$ is the general term for concentration at a distance x away from the electrode at a time t. $C_0(0,t)$, represents the concentration of ferricinium at the electrode surface and $C_r(0,t)$ represents the concentration of ferrocene at the electrode surface. The kinetics or rate of electron transfer at the electrode is determined by a combination of k^0 (which is characteristic of the analyte and the electrode) and the magnitude of the difference between E and $E^{0'}$. It is unusual for an analysis to be carried out in a region where the kinetics alone are limiting the current. Equation (3.5) contains information about whether current is limited by kinetics (represented by k^0) or mass transport of analyte [represented by the concentration of analyte at the electrode $C_r(0,t)$ and $C_0(0,t)$]. When k^0 is large or when the value of E is chosen so that the rate of reaction is sufficiently high, the mass transport limits the current. This can be controlled experimentally and under these conditions the current, usually termed the *Faradaic current,* is given by

$$i = nFAD \left.\frac{dC}{dx}\right|_{x=0} \tag{3.6}$$

where the derivative term represents the slope of the concentration profile at the electrode surface. The derivative term may be thought of as the tangent of the concentration distance plot at the electrode surface. The diffusion coefficient D is a constant characteristic of the analyte, which usually takes a value between 10^{-6} and 10^{-5} cm^2/s.

3.2.2. Practical Aspects

Before discussing the range of amperometric methods normally used, it is necessary to define the experimental conditions required for an optimum response. Points needing consideration include the need for background electrolyte, and the electrode configuration.

The reason for adding background electrolyte is to make the solution more conductive. The need for this becomes clear if we consider a twin-electrode configuration in Figure 3.3, where the two electrodes are facing each other.

If a nonpolarizable medium (e.g., pure water) is placed between the two electrodes, and a potential of 1 V is applied to one electrode with respect to the other electrode, then there will be a linear drop of potential between the two electrodes (Figure 3.3a). However, if a solution of a strong (i.e. dissociated) electrolyte is placed between the two electrodes, then the potential drop will occur across a thin layer of solution close to each electrode termed a *double layer.* The electrolyte acts in such a way as to shield this large potential (applied to the working electrode, e.g.) from the bulk of solution. Thus generally there is only a small potential field through the bulk of the solution. An advantage of this is that migration of an elec-

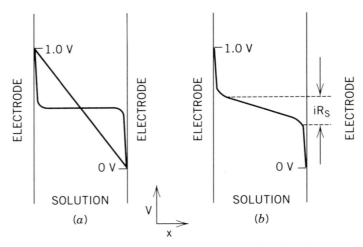

Figure 3.3. Schematic diagram of the potential drop between two electrodes. The potential drop is linear (*a*) in the absence of electrolyte. When electrolyte is added, a double layer (of nanometer-range thickness) forms at each electrode, resulting in a constant potential throughout the bulk solution. (*b*) In this case there is a potential drop through the bulk solution equal to i $\times$ R_s, where *i* is the current flowing and R_s is the solution resistance. The potential drop through the solution may be minimized by the addition of background electrolyte.

trochemically generated charged species is prevented. Migration involves the movement of a charged species under an electrical potential field and may lead to irreproducible mass transport.

If there were a resistance R_s in solution and a current *i* flowed, then there were would be a potential drop iR_s in addition the potential drop through the two double layers as may be seen in Figure 3.3*b*. The addition of electrolyte strongly reduces the resistance so that potential drop is prevented.

The presence of electrolyte results in the formation of a double layer at each electrode, which generally takes the form of ions hugging the electrode surface because of electrostatic attraction. This results in the appearance of an additional current in the outer circuit, which will interfere with the Faradaic current, related to the concentration of analyte. The problems due to double-layer charging current may be removed by the use of certain potential waveforms, as will be discussed later.

Another means of mass transport in addition to diffusion (movement of species under a concentration gradient) and migration (movement of a charged species in a potential field) is *convection,* a controlled flow of solution past the working electrode. This is typically encountered in high-performance liquid chromatography

(HPLC) or flow-injection analysis (FIA). Because of viscosity considerations, the velocity of solution drops dramatically close to the electrode surface. Thus convection, coupled with diffusion, will dictate the mass transport under these conditions.

As mentioned above, the usual configuration for electrochemical measurement is the use of a working electrode coupled with a reference electrode. So far, for simplicity, a two-electrode system has been used in this discussion. In practice, however, there are two problems with this system. The first involves the fact that any current flowing from the working electrode flows directly into the reference electrode. Whereas most reference electrodes have some tolerance for accepting current without any dramatic potential change, a point is reached where the composition of the reference electrode is changed so much by the flow of current that its potential does change. This is an unfortunate situation because if the potential of the reference or benchmark electrode changes, then any potential quoted with respect to it will be useless.

The other difficulty of the two electrode system is the presence of a resistance R_s between the working and reference electrodes. If a current i were to flow, then there would be a potential drop through the solution equal to the product $i \times R_s$. This potential drop would mean that if, say, 1.0 V were applied to the working electrode with respect to the reference, then the actual current felt by the working electrode would only be $1.0 - i \times R_s$ V.

The solution to both of these problems may be found in using a third auxiliary electrode in the cell. This electrode usually takes the form of a graphite rod or platinum mesh, which has a larger electrode area than the working electrode. The purpose of a potentiostat is to allow the application of a potential to the working electrode with respect to the reference electrode and at the same time force any current to flow into this third (auxiliary) electrode.

There is only one small problem, as can be seen in Figure 3.4a. This diagram shows the a current flux flowing between the working and auxiliary electrode. This current is passed primarily by the movement of electrolyte species through the bulk of solution. The reference electrode is sitting in this flow of current, and although there is no current flowing into the reference electrode, it still feels the current produced with the result that there still remains a potential drop equal to the product of $i \times R_s'$, where R_s' is the resistance between the working electrode and the reference electrode. This situation may be alleviated by the introduction of a Luggin capillary, which takes the form of an open glass tube hiding the reference electrode from the flux of ions in the solution as shown in Figure 3.4b. In the final case of three electrode system with a Luggin capillary, there is only a small potential drop equal the product of the current and the resistance between the working electrode and the tip of the Luggin. In this way the interfering non-Faradaic current is reduced in magnitude.

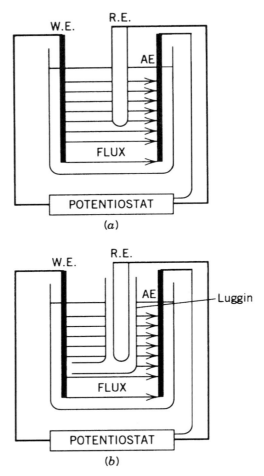

Figure 3.4. (*a*) A schematic diagram of a three-electrode cell connected to a potentiostat. The potential is applied between the working electrode and the reference electrode, and the current is forced to flow between the working and the auxiliary electrode. The potential drop between the working and reference electrode may be removed by introducing a Luggin capillary as shown in (*b*).

3.3. VOLTAMMETRIC METHODS FOR ANALYSIS

Some of the types of applied waveforms (left-hand side, Scheme 3.1) and various cell configurations (right-hand side, Scheme 3.1) have given rise to a variety of obscure nomenclature as can be seen in the centre of Scheme 3.1. If the potential is not constant, then there is a broad division into potential pulses and ramps along

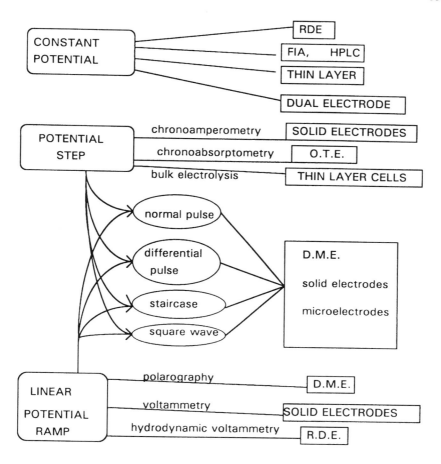

Applied potential waveforms   Cell configuratior

Scheme 3.1. An overview of the different combinations of potential waveforms (LHS) applied to the different electrode systems (RHS), giving rise to varied nomenclature. The combination of potential ramp and step techniques are hybrid and are shown in ellipses. The notation is as follows: DME; dropping-mercury electrode; FIA, flow-injection analysis; HPLC, high-performance liquid chromatography; OTE, optically transparent electrode; RDE, rotating-disk electrode.

with various combinations of both. In Scheme 3.1 the arrows linking potential step and linear potential ramp boxes lead to four combinations of methods that may be applied to the various electrode configurations shown on the right-hand side. Later there will be a discussion on why the pulsed waveform has become so popular in the area of electrochemistry, but first the simplest waveform will be dealt with.

3.3.1. Chronoamperometry

One of the simplest potential waveforms that can be applied is that of a potential step. Consider a platinum working electrode in a solution of ferrocene with appropriate background electrolyte, such as 0.1 mol/dm^3 LiClO$_4$. Initially the electrode potential applied is such that ferrocene is stable close to the electrode surface. Then a potential step is applied where the final potential is more positive than $E_1^{0'}$. As a result, the concentration of ferrocene close to the electrode falls to zero and a diffusion layer extends out into solution, as can be seen in Figure 3.2. As the slope of the concentration profile relaxes out into solution the current decays according to Eq. (3.6). An expression has been derived for the decay in current for this experimental situation, which is

$$i = \frac{nFA\sqrt{D}C_0}{(\pi t)^{1/2}} \tag{3.7}$$

This current is directly related to electrode area A, diffusion coefficient D, analyte concentration C_0 (mol/cm^3) and decays as a function of $1/\sqrt{t}$.

At short times there is an extra current due to the process of double-layer charging mentioned earlier. A characteristic of the latter current is that it decays more rapidly than the Faradaic current, as can be seen in Figure 3.5.

Thus, if the current is sampled at point 1, then the contribution to the current by the double-layer current is negligible. This point has been employed in the pulse techniques shown in Scheme 3.1.

Constant applied potential methods may be used in situations where there is a twin-electrode thin-layer configuration. For example, consider a solution of ferrocyanide [Fe(CN)$_6$]$^{4-}$ (a form of iron(II) that may be oxidized for form Fe(III), which is also a reasonably stable form of iron in solution), cholesterol oxidase (Chod), and phosphate buffer in a twin-electrode thin-layer cell as shown in Figure 3.6.

On addition of cholesterol, the following reactions occur:

$$\text{Cholesterol} + O_2 \xrightarrow{\text{Chod}} \text{Cholest-4-en-one} + H_2O_2 \tag{3.8}$$

$$2[\text{Fe(CN)}_6]^{4-} + H_2O_2 \rightarrow [\text{Fe(CN)}_6]^{3-} + 2OH^- \tag{3.9}$$

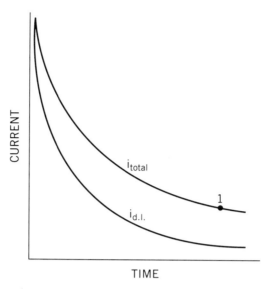

Figure 3.5. The decay in total current (i_{total}) and double-layer charging current (i_{dl}) after the application of a potential step. When the current is sampled at point 1, there is very little contribution from the double-layer charging current. This fact is applied widely in pulse techniques to achieve lower limits of detection.

Once reactions (3.8) and (3.9) are complete, there is a solution of Fe(II) and Fe(III) (Figure 3.6a). The cell is switched on at a potential where the Fe(III) is reduced at the working electrode. Meanwhile there is an equal amount of charge passed for oxidation at the auxiliary electrode forming Fe(III). Thus there is a cyclic system where the Fe(III) is being used up at the working electrode and being generated at the auxiliary electrode. This gives rise to a steady-state current, the magnitude of which depends on the original amount of cholesterol added. It takes $0.01d^2/D$ seconds for this steady-state current to be reached, where d is the distance between the electrodes and D is the diffusion coefficient of the mediator. Should this system be miniaturized to accept, for example, a drop of blood or serum, then it would readily fall into the "sensor" category since it provides a time-independent signal.

3.3.2. Linear Sweep Voltammetry and Cyclic Voltammetry

Linear sweep voltammetry involves applying a regularly changing waveform to the working electrode, as can be seen in Figure 3.7.

Once again, we take a solution of ferrocene in acetonitrile with the appropriate

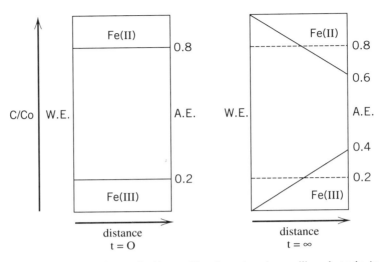

Figure 3.6. Consider a thin-layer cell with a working electrode and an auxiliary electrode. An oxidant has been added to convert some Fe(II) to Fe(III). Initially (at $t = 0$ there is a fraction (0.2) of the total concentration (C_0) of iron species in the thin layer cell converted to Fe(III). When the cell is switched on, any Fe(III) close to the working electrode is converted to Fe(II). An equal amount of oxidation [of Fe(II)] occurs at the auxiliary electrode (AE), eventually resulting in the presence of linear concentration distance profiles ($t = \infty$). At this point there will be a steady-state current related to the initial fraction of Fe(II) converted to Fe(III).

background electrolyte. Initially the potential is chosen to ensure that the ferrocene is stable at the electrode surface (E_i). Then the potential is gradually changed. The potential is scanned at a rate v V/s toward (and past) $E_1^{0'}$, the formal potential of the ferrocene/ferricinium couple. Once the potential approaches $E_1^{0'}$, any ferrocene close to the electrode is oxidized. The $E - E_1^{0'}$ term of equation (3.5) is changing in such a way as to gradually shift the equilibrium of the couple toward the ferricinium side. However, even though the potential is gradually changing, increasing the rate of reaction, there comes a point where there is no further ferrocene close to the electrode. Because of this, the current begins to decay. Thus a peak current appears when a linear potential profile is applied to a stationary electrode in solution. If the potential were held at E_λ, the current would decay (the dotted line in Figure 3.7b), according to the Cottrell equation [Eq. (3.7)]. It is possible to carry out analysis using this technique since the peak current is directly related to the concentration.

Furthermore, it is possible to apply a reverse potential ramp going in a negative direction as shown in the Figure 3.7a. In this case it is possible to capture the previously generated ferricinium ion (as long as it is stable in the electrolyte) and re-

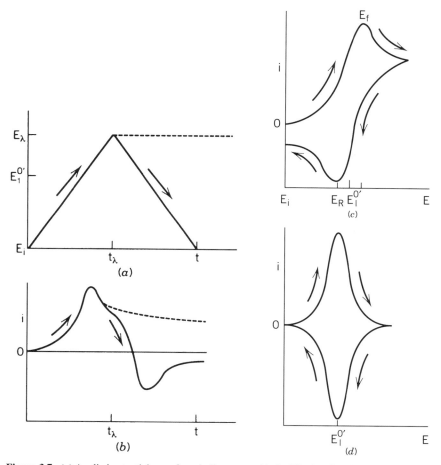

Figure 3.7. (*a*) Applied potential waveform in linear scan (dashed line) and cyclic voltammetry and the resulting current (*b*) for the linear scan (dashed line) and cyclic voltammetry (full line). Plots (*a*) and (*b*) represent a function of time, and the initial oxidation current (due to the oxidation of fc) is followed by a reduction current (when the fc⁺) is reduced. When the current is plotted against the applied potential curve, plot (*c*) results. (*d*) Cyclic voltammogram when the waveform in plot (*a*) is applied to a solution in a thin-layer cell. The current returns to zero after the peak, indicating bulk electrolysis.

duce it by sweeping the electrode potential back past the $E_1^{0'}$ to a potential region where the ferrocene is again stable at the electrode. Thus a reduction peak is obtained. When the current is plotted against the applied potential, a curve as in Figure 3.7*c* is obtained. The $E_1^{0'}$ is to be found as half-way between the forward (E_f) and backward (E_r) peak potentials. It is possible to gain important information about the couple by carrying out cyclic voltammetry. Once again there is a problem with double-layer charging current; a current which is present because the electrode potential is gradually changing.

In the case of thin-layer cells, where there is a finite amount of solution, a more symmetric cyclic voltammogram is obtained on application of a cyclic ramped waveform as may be seen in Figure 3.7d. In this case the positions of the forward and reverse peaks are the same and is equal to $E^{0\prime}$ of the couple. On the forward scan the current decays to zero because all the material in the thin layer is oxidized and it amounts to a bulk electrolysis. The expression for peak current is related to concentration as follows:

$$i_p = \frac{n^2 F^2 V C_0 \upsilon}{4RT} \tag{3.10}$$

where V is the volume of the cell, C_0 is the concentration of the species in the cell, and the other terms have their usual meaning. Cyclic voltammetry of thin coatings containing redox species results in plots similar to that in Figure 3.7d since they behave like a thin-layer cell.

3.3.3. Rotating-Disk Electrochemistry

This technique consists of a linear potential ramp waveform applied to a disk electrode that is rotating at a rate of ω s^{-1}, where $\omega = 2\,\pi RPM/(60)$, where RPM is the rotation rate in revolutions per minute.

As can be seen in Figure 3.8, when the potential is increased toward $E_1^{0\prime}$, the current increases accordingly. However, instead of decaying subsequently as in the case of voltammetry, the current reaches a limiting value. This happens because of convective mass transport where reagent is being continuously being swept toward the electrode. The limiting or steady-state current is directly related to the rotation rate and the concentration of electroactive species in solution. Although not too often used in analysis, the main use of the rotating-disk electrode (RDE) is to measure fast heterogeneous rate constants.

3.3.4. DC Polarography

This method consists of a linear potential ramp applied to a dropping-mercury electrode (DME). The direct-current (DC) response is sigmoidal as in the case of rotating-disk electrochemistry (Figure 3.8) since the solution is continuously stirred by the appearance of new drops at a regular time interval. Historically it is an important method for which Heyrovsky achieved the Nobel prize. There are several advantages with the use of mercury for trace-metal analysis. First, as each drop appears, the electrode surface is new and well defined, unlike at solid elec-

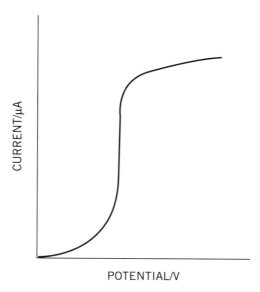

Figure 3.8. The current–potential behavior when a linear potential scan is applied to a rotating-disk electrode in a solution of a reversible species. In some cases it is difficult to evaluate the limiting current because of the sloping baseline.

trodes, which may contain surface oxides and that frequently need to be pretreated before use. When trace metals are reduced and deposited on a solid electrode, a nucleation step is required. However, for a range of heavy-metal cations such as Pb^{2+}, Cd^{2+}, and Cu^{2+}, the reduced form of the metal dissolves in the mercury forming an amalgam. This is a much simpler process, and more reproducible current–potential profiles are obtained. A further advantage is that it is difficult to reduce H^+ at mercury (compared to some solid electrodes) and this makes the analysis of metals in acid solution possible. Quantitative determination of a number of metals simultaneously may be carried out as shown in Figure 3.9. Their concentrations are directly related to their limiting currents. The metals may also be identified by their half-wave potentials; the value of potential at a current equal to half the limiting current.

The method is quite complicated and does not fall into the simple "sensor" category. The use of mercury as an electrode material requires great care because of its reasonably high vapour pressure. Other problems include clogging of the capillary along with the presence of nonfaradaic "peak maxima" currents. Furthermore, as each drop appears into solution, a new double layer has to be formed. This results in a high residual current that limits the detection limit of the technique. Finally, there is the problem associated with measuring the limiting currents at low

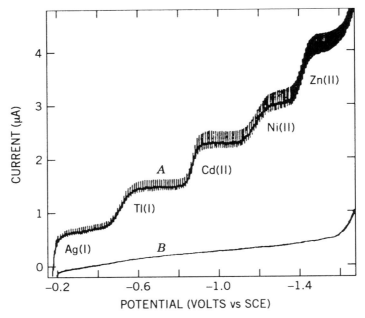

Figure 3.9. Polarograms of (curve *A*) approximately 1×10^{-4} mol/dm³, each of Ag⁺, Tl⁺, Cd²⁺, Ni²⁺, and Zn²⁺, listed in the order in which they appear (the solution was in aqueous 1 mol/dm³ NH₃, 1 mol/dm³ NH₄Cl containing 0.002% Triton X-100 and (curve *B*) the supporting electrolyte alone. (Reprinted with permission from L. Meites, *Polarographic Techniques*, 2nd ed., Wiley, New York, 1967, pp. 164.)

analyte concentrations and sloping baselines. The original technique has been modified by the use of more complicated pulse waveforms to lower the detection limit.

3.3.5. Pulse Methods

As mentioned above, there is a large residual current in polarography due to the new double layer required for each drop. The purpose of these pulse methods is to discriminate against the double-layer charging current. This is done by taking advantage of the fact that the timescale to form the double layer is much shorter than that of the Faradaic current, as may be seen in Figure 3.5. Furthermore, as each drop grows, the Faradaic current gets larger as a result of the larger surface area.

Normal pulse polarography involves applying a more complicated waveform than a linear ramp. The applied potential wave form is shown in Figure 3.10*a*.

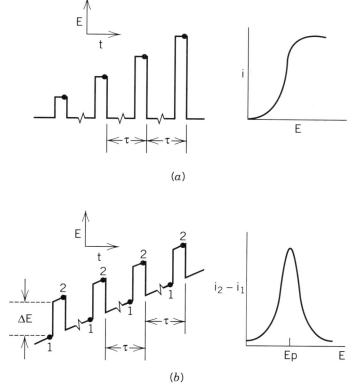

Figure 3.10. (*a*) Normal pulse polarography, where τ represents the drop lifetime. The current is sampled at the position of the filled circle and a sigmoidal *i/E* plot results. (*b*) Differential pulse polarography, where the pulses are applied in sequence with the drop lifetime of length τ. Potential pulses (ΔE) are applied to a slowly increasing ramp, and the difference between the current sampled at the point 2 and the point 1 is output in for each drop. The result is a peaked waveform.

The pulses are applied in phase with the drop lifetime. The current is sampled at the point indicated by a closed circle long enough after the pulse is applied to allow the double-layer charging to decay and the Faradaic contribution to dominate. Furthermore, the current is sampled at the end of the drop lifetime when the drop area is greatest, unlike in polarography, where the current is monitored continuously giving rise to noise that has to be filtered out. In addition to its use with a dropping-mercury electrode, this waveform has been employed at microelectrodes for analysis in vivo because of the wide potential excursions that serve to clean the electrode from interference due to adsorbed organic species.

Differential pulse polarography (DPP) involves the application of a series of pulses to a slowly linearly increasing ramp as can be seen in Figure 3.10b. Once again, these pulses are in phase with the drop lifetime and the current is sampled at two points on the waveform: once just before the pulse is applied and again well after the pulse is applied. For a dropping-mercury electrode, this means that there will be a signal in a region where the total current is changing rapidly. The output of this technique will be zero before the wave and at the plateau of a conventional polarogram and takes the form of a symmetric peak which is more useful from an analytical point of view. The advantage of these pulse methods is that there is little double-layer charging contribution to the response, which results in a lower detection limit. Whereas in polarography the detection limits are 10^{-6}–10^{-5} mol/dm^3, for DPP the detection limits can be 10^{-8} mol/dm^3 depending on the metal cation analyzed.

Staircase and square-wave voltammetry techniques have been used widely for solid electrodes. The staircase voltammetric method involves the application of a digitized linear ramp. With digital data acquisition techniques, staircase voltammetry is more sensible than aiming for a linear ramp. The advantage over cyclic voltammetry is that the current is sampled a little later than when the step is applied, which allows discrimination over double-layer charging, as may be seen in Figure 3.11a. Under conditions where the pulse width and the potential step height are small, the current response for staircase voltammetry is identical to linear scan voltammetry. This technique has been used for analysis in flowing streams and in liquid chromatography.

Square-wave voltammetry is an allied technique where the waveform is as shown in Figure 3.11b. Currents can be sampled at points 1 and 2, and either these or their difference can be used as the response. Note that rather than being a ramp-based technique, it is closer to a digitized step (Figure 3.11a), with increasing and decreasing pulses. The technique has been used at small solid electrodes and when used at a DME, the plateau current due to oxygen reduction is subtracted out. Square-wave voltammetry is one of the fastest techniques compared to normal pulse and differential pulse methods. In fact the area of applying complex waveforms under computer control has progressed to the extent that most problems with these techniques are due to mercury delivery and cell design.

3.3.6. Stripping Methods

These methods involve a preconcentration step before analysis and are designed to achieve a lower detection limit. The technique may take the form of accumulation of metal as amalgams into a hanging mercury drop or a mercury film. Subsequently the accumulated metal may be stripped out by applying one of the waveforms mentioned above. Such a technique is termed *anodic stripping voltammetry*.

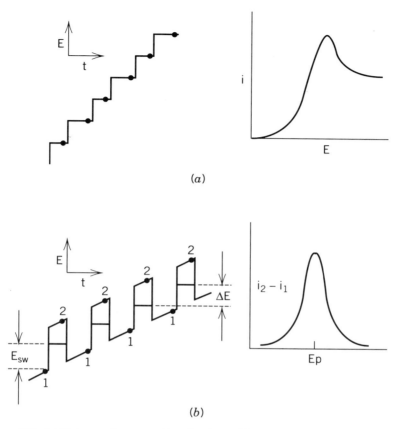

(a)

(b)

Figure 3.11. (a) Staircase voltammetry. Note that the resulting current–potential profile is not symmetric as in the case of linear scan voltammetry. (b) Square-wave voltammetry; E_{sw} is the applied pulse amplitude (up and down) to a staircase voltammetric waveform, where ΔE is the regularly increasing step in potential. The difference between the current sampled at point 2 and point 1 results in a symmetric waveform.

Rather than using mercury, it is also possible to accumulate the analyte in a coating on the electrode surface. This is achieved by depositing a polycationic (e.g., quaternized polyvinylpyridine) or polyanionic (polystyrenesulphonate or Nafion) layer on a solid electrode. Alternatively, adsorption of particular compounds may be used to accumulate them on electrode surfaces, following which they can be oxidized. Using stripping methods, very low concentrations of amalgam forming metals may be detected, as low as 10^{-10} mol/dm^3.

3.4. ELECTROANALYSIS IN FLOWING STREAMS

Electrochemical detectors in flowing streams such as in flow-injection analysis (FIA) or high-performance liquid chromatography (HPLC), may be used to detect a wider range of compounds than UV–visible methods. Flow-injection analysis involves the introduction of a sample into a carrier of reagent, which allows the analysis of a certain analyte in the plug. It is a very simple technique amenable to portable instruments. The cell configuration may take the from of a thin layer or a wall-jet, where the effluent from the column impinges perpendicularly onto the surface of the working electrode.

In general, the thin layer planar electrode configuration is operated at a constant applied potential, which means that there should be no double-layer charging current. As a rule, flow-through detectors should be easy to disassemble and should not leak. The position of the counter and reference electrodes relative to the working electrode is also important. More recently coated electrodes have been used in flowing streams to enhance selectivity. For example the negatively charged ionomer Nafion may be used to exclude anions and selectively include cations from a flowing stream.

In the wall-jet configuration the outlet of the column impinges perpendicularly onto an electrode surface. This electrode configuration depends greatly on the nozzle diameter and its distance from the working electrode. If the electrode area is much greater than the nozzle diameter, the sensitivity is much greater. Steroids have been detected using a thin-layer wall-jet electrode configuration. Also, copper has been determined in urine at microelectrodes using a wall-jet configuration.

Reticulated vitreous carbon is a fine honeycomb form of carbon that, when filled with epoxy, allows thin lines of carbon to be exposed to the solution that behave like an array of microelectrodes. This type of electrode has been shown to yield an enhanced signal-to-noise ratio compared to planar electrodes when used in a FIA configuration.

Dual-electrode detectors take the form of two embedded electrodes in a thin layer either parallel or across the flow of solution. In a parallel configuration it is possible to have a generator–detector system, whereas if the two electrodes are beside each other, it is possible to coat one and look at the difference in response between them. Reagent generation has been used for dual electrodes in series where Br_2 has been generated to detect phenolic ethers.

Pulsed amperometric detection (PAD) involves the use of cleaning desorption waveforms on a regular basis as shown in Figure 3.12, where E_2 and E_3 are used for cleaning, and the signal is accumulated at E_1. The current may be sampled (a), or integrated (b), or integrated over a small cyclic ramp as shown in Figure 3.12c. Species may be determined at bare electrodes (alcohols) or else by adsorption at oxide coated electrodes (aliphatic amines, or sulphur compounds). The current associated with the oxidation of the metal may interfere with the analyte signal in

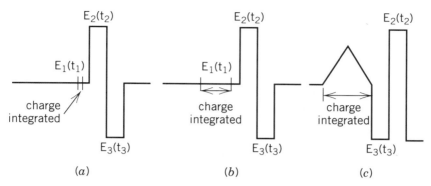

Figure 3.12. (*a*) Pulsed amperometric detection (PAD). The current is sampled toward the end of E_1. (*b*) The current is integrated over a short time at E_1. (*c*) The current is integrated while the cyclic scan is being applied. Time t_1 in (*a*) = 16 ms, t_1 in (*b*) and (*c*) = 200 ms; t_2 = 50–200 ms, t_3 = 200–400 ms in each case.

the latter case. The above two methods depend on the choice of E_2. This waveform can be applied at a sufficiently high frequency (1–2 Hz) to be used as a detector in HPLC. The waveform in Figure 3.12*c* is termed *integrated pulsed amperometric detection* (IPAD), and an example of its use may be seen in Figure 3.13, where three amino acids are detected at a gold electrode. It can be seen that the response using IPAD (curve 3.13*b*), is more sensitive than that of PAD (curve 3.13*a*). PAD and IPAD are useful means of detection in HPLC where UV–vis spectroscopy is inadequate.

3.5. MICROELECTRODES

3.5.1. Voltammetry at Microelectrodes

Microelectrodes are working electrodes that are constructed to have a very small surface area. They can be made by embedding a platinum wire (1, 5, or 10 μm) or carbon fiber (8 μm) in glass or epoxy and exposing the cross-sectional disk to solution. Thus, their electrochemical area is tiny, and they can be used in a two-electrode configuration with conventional-sized reference electrodes. In this case the amount of current is so small that it will not change the composition of the reference electrode and there will be no significant potential drop through solution. Another advantage associated with a microelectrode compared to its larger area counterpart, and that is due to the "edge effect," where there is a greater Faradaic:double-layer charging current ratio than at a macroelectrode. Difficulties asso-

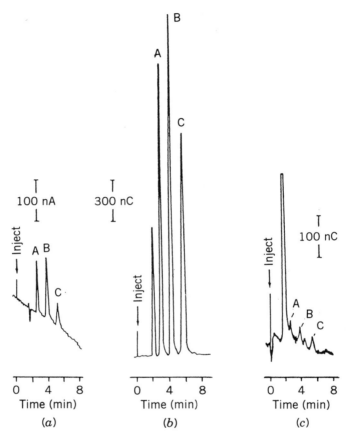

Figure 3.13. Isocratic separation of three amino acids. Electrode was gold with SCE reference. Column: Dionex HPIC-AS6. Mobile phase: 0.10 mol/dm^3 NaOH. Detection: (*a*) PAD; (*b*), (*c*) IPAD. Peaks *A*, lysine; *B*, aspragine; *C*, 4-hydroxyproline. Concentrations: (*a*), (*b*) ~50 ng each, (*c*) ~0.5 ng each (i.e., ~3 pmal). [Reprinted with permission from D. C. Johnson and W. R. LaCourse, *Anal. Chem.* **62**, 589A (1990).]

ciated with the microelectrode include the requirement for a low noise means of current measurement along with practical problems of polishing and manipulation.

When a waveform as shown in Figure 3.7*a* is applied to a microelectrode, the current response is quite different from that at a macroelectrode. The current increases as the reaction is forced to the right-hand side, but it then reaches a steady state current. As the diffusion layer expands out into solution (Figure 3.14*a*), it adopts a hemispherical shape, where its surface area is gradually increasing. Thus there is an ever-increasing catchment area for reagent, and this compensates for

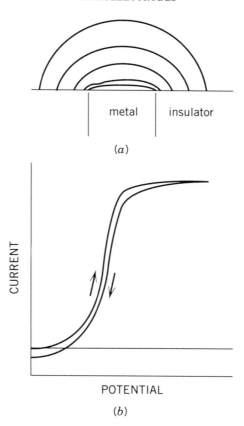

metal   insulator

(a)

CURRENT

POTENTIAL

(b)

Figure 3.14. When a potential waveform shown in Figure 3.7a is applied to a microdisk electrode, a steady-state current is obtained and a sigmoidal i/E curve is obtained. (a) A plot of the expanding hemispherical diffusion layer which gives rise to the steady-state current; (b) a typical experimental curve exhibits some hysteresis, which may be minimized experimentally.

the decreasing concentration distance slopes at the electrode (Figure 3.2). The value for the steady-state current, the plateau in Figure 31.4b, is given by the following equation:

$$i_{ss} = 4nFC_0Dr \tag{3.11}$$

where r is the radius of the electrode disk and D is the diffusion coefficient for the reagent species.

There have been unusual applications of microelectrodes. For example, consider the use of microelectrode as a detector in supercritical CO_2, saturated with water but not containing any background electrolyte for the determination of fer-rocene, even though the solvent has a very high resistance. Also electrochemistry has even been carried out in the gas phase for detecting eluents from a gas chro-matograph. This latter process occurs possibly as a result of surface conductivity across epoxy between the two concentric ring microelectrodes. This configuration is ideal in terms of a simple sensing configuration. Rather than have a complicat-ed three-electrode cell, it is possible to have a simple two-electrode arrangement that may be subsequently coated with layers that impart some specificity to the sensing process. Finally the use of microelectrodes for solid-state electrolytes has been employed where a 10-μm platinum microdisk electrode has been coated with a crosslinked polyethylene oxide doped with $LiClO_4$. The electrochemistry of 100-μm-thick layers of this solid polyelectrolyte containing ferrocene derivatives has been examined as a function of temperature. It is possible that some forms of this type of polyelectrolyte may be used in a gas-sensing configuration where the pres-ence of a gas solubilized in a solid electrolyte may enhance the conductivity with-in the polyelectrolyte containing an electroactive reagent yielding a response. Rather than having electrochemistry in only a liquid phase, it is possible with mi-croelectrodes to extend the process to the gas and solid phases.

3.5.2. Microelectrodes in Flowing Streams

Microelectrodes possess considerable ability for enhanced Faradaic current. Fur-thermore, they draw only a small amount of current, which makes them amenable to two-electrode configurations. Also they can be used in situations where there is a low background electrolyte concentration. They have been used in in vivo analy-sis, and for heavy-metal analysis. They can take the form of microdisks, micro-rings, and microcylinders composed of carbon fibers used in solution. Fast scan voltammetry has been carried out at microelectrodes, and it is possible to measure fast heterogeneous rates of reaction using this method.

Arrays of microelectrodes take advantage of the enhanced diffusion as well as providing a greater (and more easily measurable) current. They have also been used as detectors in HPLC. The use of carbon fibers in flowing streams increases the signal to background ratio by as much as a factor of 1.6. Hydroquinone con-centrations as low as 1.1×10^{-7} g/dm^3 have been determined in FIA using mi-croelectrodes.

3.5.3. Microelectrodes in Clinical Chemistry

Microelectrodes have been popular for spatial characterisation, especially in the neurochemical sciences. These have been reviewed as long ago as 1976, and tung-sten microelectrodes have been used in vivo as far back as 1957. There are many problems associated with the intrusion of a microelectrode into a biological sys-tem since the latter usually reacts to protect itself.

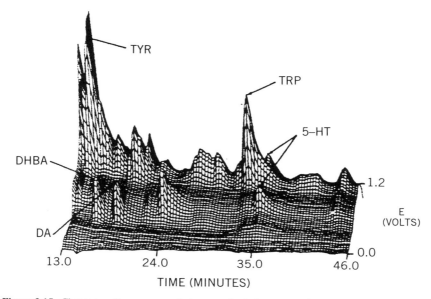

Figure 3.15. Chromatovoltammogram of a homogenized giant neuron from a land snail *Helix axpersa* using a 19-μm-i.d. HPLC open tubular column coated with dimethyloctadecylsilane with a 9-μm carbon fiber inserted into the outlet of the column as detector (DA, dopamine; DHBA, 3,4-dihydroxybenzylamine; 5-HT, serotonin; TYR, tyrosine; TRP, tryptophan). The potential was ramped from 0.0 to 1.3 V at 1 V/s in each case, and the background was subtracted. Mobile phase was 0.1 mol/dm³ citrate buffer adjusted to pH 3.1 with NaOH and also contained 2.1×10^{-4} mol/dm³ dimethyloctylamine and $6.2 - 9.5 \times 10^{-4}$ mol/dm³ sodium cetyl sulfate. [Reprinted with permission from R. T. Kennedy and J. W. Jorgenson, *Anal. Chem.* **61**, 436 (1989).]

The concept of spatial characterization has been quantified in the use of a *scanning electrochemical microscopy* (SECM), also termed *scanning electrode microscope* (SEM), where a microelectrode is rastered across the surface of a conductive material and that allows vertical resolution of tens of nanometers. Using this technique, the electrochemical conditioning of a carbon electrode has been spatially characterized.

Microelectrodes have been used for anodic stripping voltammetry, and diffusional enhancement has been observed and it is possible to apply very fast scan rates, thus reducing the timescale of the experiment. When mercury is used as a macroelectrode, deoxygenation is frequently the slowest step in the process. However, backgrounds carried out in air-saturated solutions have been subtracted from the total response using square-wave methods at microelectrodes.

Individual carbon fibers have been used as voltammetric detectors in open tubular capillary columns, and using fast linear scan rates, it is possible to develop a chromatovoltammogram, a plot of time since injection versus applied potential

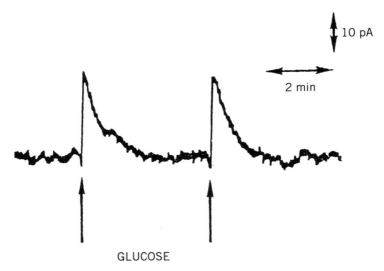

GLUCOSE

Figure 3.16. Amperometric current monitored at a 2-μm glucose electrode placed in the cyctoplasm of the large dopamine neuron of the pond snail. This neuron has a diameter of ~200 μm and a volume of 4–6 nL. The arrow indicates two successive injections (each 2 pL) of 3 mol/dm³ glucose solution into the cytoplasm. Injections were carried out with an Eppendorf pressure-based microinjector system and a glass micropipette having an approximate diameter of 1 μm. The electrode contained glucose oxidase crosslinked with gluteraldehyde. [Reprinted with permission from T. Abe, Y. Y. Lau, and A. G. Ewing, *J. Am. Chem. Soc.* **113,** 742 (1991).]

versus current due to redox reactions of eluent species such as in Figure 3.15. In these experiments the column has an internal diameter ranging from 10 to 25 μm and a total volume of 390 nL. Typical injection volumes were 5 nL. Analysis for particular metabolites within cells has been accomplished, by homogenizing the cell and after centrifugation, a sample of the supernatant is injected onto the column. In the figure the peak for tryptophan corresponds to approximately 18.2 ± 7 fmol of material.

The use of an array of 16 microelectrodes each held at a specified potential or scanned over a limited potential range has also been used to generate a chromato-voltammogram. These can produce current–potential responses over very short timescales with virtually no double-layer charging. Coatings have also been applied to arrays of electrodes held at similar potentials, and information theory can be applied in order to obtain analytical information.

Pushing back the boundaries even further, 2-pL injections of 3 mol/dm³ glucose into a single large dopamine cell of a pond snail have been detected using a 2-μm glucose electrode, as can be seen in Figure 3.16. Advances such as this may ultimately lead to a glucose monitoring system linked to a glucose delivery system in vivo.

3.6. MODIFIED ELECTRODES AS AMPEROMETRIC SENSORS

3.6.1. Introduction

One inherent problem in electroanalysis using amperometric techniques is that electrochemical reactions are very dependent on the type of electrode used. For many redox species, electrochemical reactions are occurring at high overpotentials, or are sometimes not observed at all, even though they are thermodynamically possible. Such problems can often be traced to fouling of the electrode by adsorption or by chemical reactions involving the electrode surface. It is for these reasons that amperometric sensors have had a limited applicability in the past.

However, the fact that the nature of the electrode–electrolyte interface is important for the actual electrode reaction taking place can be used to our advantage by tailoring this surface by changing the nature of the electrode–electrolyte interface. The well-known carbon paste and mercury film electrode are the first examples where surfaces other than the traditional mercury drop, solid metal, or carbon surfaces were used.

In recent years much research has been carried out in the area of modified electrodes. By developing the modified electrode, one attempts to exert greater control over the electrochemical reaction by the application of a thin layer onto conventional solid electrode. In principle, a very selective electrode can be obtained. By the application of a layer appropriate for a particular analysis, a "designer electrode" can be developed. So far this approach has been most successful with the development of amperometric biosensors based on immobilized enzymes. One could consider modified electrodes to be the amperometric analogs of the ion-selective electrode.

Of particular interest is the application of electrocatalytic layers that actively participate in the electrochemical processes. In modified electrodes coated with electrocatalytic layers the analytically important redox process takes place between the coating applied, and the substrate and the underlying electrode is solely responsible for the source or drain of electrons. In this manner the electrochemical properties of the electrode are controlled by the modifying layer and not by the underlying carbon or metal surface. Coatings that are actively involved in the electrochemical process are called *mediating layers*. In principle, a whole range of modifying species is available, these vary from metal deposits to organic layers, polymers, and immobilized enzymes. The choice of coatings is determined mostly by attempts to activate or passivate particular reactions. Attempts have also been made to protect the electrode surface from unwanted reactants and so avoid fouling of the electrode surface. In the last case the layers applied often do not have mediating properties but simply protect the electrode surface from unwanted species.

In the next section (3.6.2) the preparation of modified electrodes will be discussed first, outlining modification methods and materials used. Then the operation mode of modified electrodes will be discussed with particular emphasis on the mediation process as it is observed for electroactive coatings, including some examples of the application of modified electrodes as sensors. In Section 3.7 attention is given to biosensors based on electrode modification methods. This section does not aim to give a complete literature survey of the area of modified electrodes. For more detailed discussions on modified electrodes the reader is referred to recent reviews by Hillman, Murray, and others.

3.6.2. Electrode Preparation

The methods and materials used to prepare modified electrodes are often directly related to the particular problem to be solved. However, there are some common principles that have to be considered to obtain a useful modified electrode. It is very important that electrical contact be established between the analyte and the underlying solid electrode. For a mediating electroactive layer, electrons should be able to move through the modifying layer to or from the electrode surface depending on whether an oxidation or reduction process is taking place. In other words, charge transport through the layer should be efficient. When the modifying layer is simply present to protect the underlying electrode from unwanted solution species, then the layer has to be permeable to the analyte of interest, and efficient transport of the analyte through the modifying layer to the underlying electrode must take place. These conditions greatly limit the thickness of the layer that can be applied. While a very thick mediating layer would seem initially very efficient, as many redox active centers are available for mediation, thick layers often have bad charge-transport characteristics and as a result sensitivity is reduced. So it is important to determine an optimium layer thickness. The authors have found that with osmium- and ruthenium-based polymers of the type $[Os(bpy)_2(PVP)_{10}Cl]Cl$ (for structure, see Figure 3.17), where bpy = 2,2'-bipyridyl and PVP = poly-4-vinylpyridine maximum analytical response is obtained for layers of about 1 μm thickness or less. However, the optimum thickness can vary from one coating material to the other and with every application.

Another factor to be considered is the stability of the modifying layers. There are two types of stability to be considered: the chemical stability and the physical stability. One has to consider whether layers retain their essential chemical composition for a sufficiently long time but also whether the layers remain firmly attached to the electrode surface. The last factor is obviously very important, and under severe hydrodynamic conditions, as for example, encountered in flow-injection analysis, it can be quite difficult to achieve sufficient adhesion. Similar problems have been observed with modified rotating-disk electrodes. This is specially so when preformed polymers are used. In order to obtain an optimum response, electrolytes are often used that partly solubilize the polymer and this has obvious implications for the adhesion of the layer. While the physical stability of

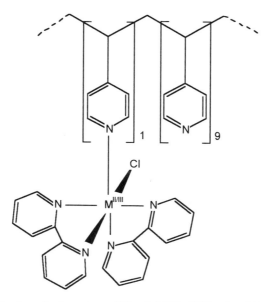

Figure 3.17. Structure of redox polymer $[M(bpy)_2(PVP)_{10}Cl]Cl$, where M is Ru(II) or Os(II).

the layers can be hard to control, the electrochemical stability can be predicted before the layer is applied, for example, by solution-based electrochemical studies of the redox active species or its mononuclear analogues.

Taking into account these considerations, a number of approaches can be made to modify a solid electrode surface with a mediating layer. The most important methods used for the modification of electrode surfaces can be summarized as follows:

1. Covalent attachment by either silanization or direct bonding
2. Polymer coatings
3. Modified carbon paste electrodes

Covalent attachment, particularly silanization, was one of the first methods used to modify electrode surfaces. In the covalent attachment method the normally electroactive species is directly bound to the electrode surface by chemical means. The electrode surface is treated in such a way that active groups are formed and these are subsequently reacted with the appropriate redox active group. Because of the nature of this method normally monolayers are obtained, although the preparation of multilayers is possible. In the silanisation process a derivatized redox active group containing trialkoxy- or trichlorosilanes is reacted with hydroxy or oxide groups present on the surface of the electrode and is, as a result, directly bound to the electrode surface. Silanization reactions have been carried out on electrode materials such as carbon, platinum, and several metal oxides. Reactive groups such as carboxylic acids can be introduced on a solid electrode surface by a variety of methods. The most popular of these include thermal and radio-frequency plasma treatment.

With the direct attachment method, usually only monolayers can be applied. These layers have limited application because of stability problems. By using polymeric coatings the electrodes can be coated with much thicker layers and great flexibility exists in the choice of the coating material. The application of polymeric layers is also more convenient and very easily carried out. In general, multilayers offer many advantages over monolayers, because the surface coverage, and therefore the amount of electroactive material attached to the surface, can be greatly increased. In addition, multilayers combine the advantages of the monolayer modified electrodes with those of homogeneous catalytic systems. Like monolayers, multilayers offer a high localized concentration of catalytic sites with an easy separation of reaction products from the catalyst. However, since redox catalysis may occur because of structural as well as chemical reasons, a three-dimensional arrangement of catalytic sites might have additional advantages. Polymeric coatings can be applied using two different methods, either by electropolymerization or by solution casting of preformed polymers. Electropolymerization is carried out by placing the electrode to be modified in a solution containing the monomer of interest and keeping the potential at the appropriate value or by cycling the potential over an appropriate range. In this manner coatings based on conducting polymers such as polypyrrole and polyaniline are obtained. During this electropolymerization, pyrrole or aniline is oxidized and a film is formed in a very controlled manner incorporating a negatively charged ion that acts as a charge-compensating ion. The advantage of this method is that, by control of the amount of current, the thickness and to a certain extend also the structure of the materials can be controlled quite well. Figure 3.18 shows the electropolymerization of polyaniline. It can be seen clearly that when the electrode is being held at the appropriate potential, the height of the cyclic voltammetry peaks increases with time. This shows that with time, and therefore with increasing amount of current, the amount of material on the electrode increases. The thickness of the layers obtained are therefore well defined and in general smooth, reproducible structures are obtained. This is shown in Figure 3.19 for an electrode coated by the anodic electropolymerization of quaternized 3-(pyrrol-1-yl-methyl)pyridine. The scanning electron micrograph shows a very compact structure with a uniform layer thickness. The structure of the polymer layers obtained by electropolymerization can be influenced to a certain extent by the choice of the counterion, and sensor applications of these materials are often very much determined by the charge-compensation counterion used. Biosensors have, for example, been prepared in this manner by

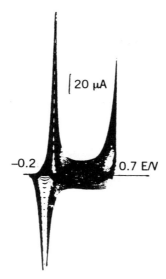

Figure 3.18. Growth of polyaniline film from 0.5 mol/dm³ aniline solution in NH₄F + 2.3HF using cyclic voltammetric scanning between −0.2 and 0.68 V. [From E. M. Genies and M. Lapkowsky, *J. Electroanal. Chem.* **326,** 189 (1987).]

the incorporation of enzymes and antibodies as counter ions. Polymer films can also be obtained by electropolymerization using reduction processes, in this case, by using redox active metal complexes that contain polymerizable groups such as vinyl groups. A disadvantage of this approach is that the films obtained can be heavily crosslinked and compact. This limits their application as sensor materials as electrolyte and analyte access can be difficult. The method is furthermore limited because it relies on the fact that the appropriate monomer can be prepared and that electropolymerization is efficient.

Another method that has been used effectively is the casting of films from solution using preformed polymers. This technique offers a wide range of materials, and using well-known synthetic techniques, systematic variations in the structure of these materials can be introduced. Different approaches can be followed to introduce the electroactive center in the polymer coating. The electroactive center can, for example, be incorporated after polymer deposition on the electrode surface. This can be achieved by an ion-exchange process. In this case the electroactive species is introduced into the layer when an electrode coated with an ion-exchange material is exposed to a solution containing the redox active species. The exposure times used for such absorption processes can vary from several minutes to hours. In this manner electroactive compounds can be introduced to ion-ex-

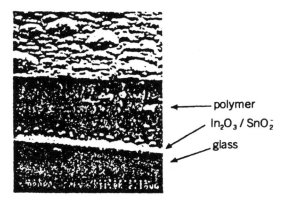

Figure 3.19. Scanning electron micrograph of a poly 3-(pyrrol-1-yl-methyl)pyridine film on indium/tinoxide-coated glass slides. [From H. Mao and P. G. Pickup, *J. Electroanal. Chem.* **265,** 127 (1989).]

change materials such as Nafion (for structure, see Figure 3.20) and quaternized polyvinylpyridines. Electroactive species that have been introduced in this way include iridates and ferro- and ferri-cyano compounds. As the binding in this case is electrostatic by nature, subsequent leaching out of the electroactive agent constitutes a problem. Coatings prepared in this manner therefore often have a limited stability.

An increased stability can be obtained when the electroactive center is covalently bound to the polymer backbone. The advantage of using a preformed polymer is that the coating material can be characterized before application with regard to molecular mass, redox behavior, and spectroscopic properties. This approach also gives one the opportunity to change to polymer backbone without altering the electroactive site and also to change the concentration of the electroactive site in the coating. An example of such a material is the metallopolymer shown in Figure 3.17. For this material the amount of electroactive osmium groups is given by n, the number of pyridine groups per osmium center. By experimental procedures, the value of n can be varied ($n = 5$, 10, 20, 50, etc.). In this manner the electrochemical characteristics of the coatings can be manipulated quite effectively. Polymeric coatings can be applied to a wide range of electrode materials, including graphite, glassy carbon, gold, and platinum. The adhesion of the layers does, however, change from material to material, and the surface treatment of the electrode surface before coating can be crucial. The mechanism of adsorption of polymer layers is at present not well understood but most likely is controlled by nonspecific adsorption and by the solubility of the polymer in the test solution. It is found that in general high-molecular-weight materials show better adhesion than

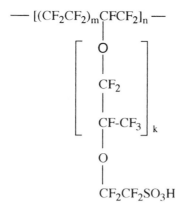

Figure 3.20. Structure of Nafion.

do low-molecular-weight polymers. To obtain stable coatings, it is therefore important to select polymers with a high molecular weight. The stability can be further increased by crosslinking methods, however extensive crosslinking can impede ion and analyte transport and is to be avoided. Films can be applied by a variety of methods such as solution casting, dipcoating, and spin coating. Using solution casting, the amount of material on the electrode surface can be controlled by the concentration and the amount of the polymer solution applied. The actual layer thickness is, however, less well defined than when electropolymerization techniques are used. This is important as often the analytical signal obtained with these modified electrodes will depend on the thickness of the modifying layer. This variation in layer thickness is caused not only by the fact that the amount of material transferred to the electrode surface cannot be controlled as accurately as for electropolymerization but also by the fact that preformed polymers tend to have a much stronger interaction with the contacting electrolyte than do electropolymerized materials. As a result, the swelling of such layers is much more pronounced and at the same time difficult to quantify. There is, therefore, not necessarily a well-defined relationship between the layer thickness that can be calculated from the amount of material applied and the actual layer thickness when the coating is in contact with a test solution.

An alternative approach is to embed, for example, electroactive or ion-exchange materials into a carbon paste matrix. For modified carbon paste electrodes, the charge transport is fast and at the same time a large surface area can be obtained. One advantage of the carbon paste electrode is that the surface can very easily be renewed and the amount of modifier can be controlled. With this method of modifications there are no problems with the adhesion of the coating. There are, however, problems with the introduction of such electrodes into FIA systems. These have been overcome by using the carbon–epoxy method. In this method the carbon paste is mixed with an appropriate epoxy and after treatment a hard, polishable surface is obtained. This method also gives improved stability and a good reproducibility.

The preceding examples show that there exists a wide range of methods and materials for the modification of electrode surfaces. This is of great importance for the development of new selective sensors. It is anticipated that by using the appropriate synthetic conditions, the specificity and accuracy of the sensors can be fine-tuned. There are, therefore, many more ways to control the electrochemical processes using modified electrodes than with the traditional electrode materials. In the next section a number of modes of operation for modified electrodes will be discussed.

3.6.3.   Operation Modes

3.6.3.1.   Preconcentration

The concept of preconcentrating analytes at electrode surfaces is not new. The three-dimensional nature of the mercury drop held at an appropriate potential allows for the effective electrolytic scavenging of analyte from very dilute solutions to produce high surface concentrations of analyte. When the potential is swept in the opposite direction, a large analytical signal may be obtained that greatly enhances sensitivity and lowers the limit of detection (LOD). With mercury electrodes this technique is only applicable to reversible redox couples such as heavy-metal ions over a limited potential region. Using modified electrodes, the potential range can be broadened considerably and irreversible electrode reactions can be exploited. The principles of ion exchange, complexometry, and precipitation have been used to preconcentrate analytes at the surface of modified electrodes. Chemically reactive sites in polymeric films have also been used to effect preconcentration. In addition to the sensitivity enhancement that can be obtained, careful choice of the electrode modifier can enhance selectivity on the basis of electrostatics, chemical functionality, stereochemistry, selective binding, or molecular size. In order to obtain maximum benefit from preconcentration, this process should take place in a three-dimensional region close to the electrode surface. For this reason polymeric films have predominated preconcentration at modified electrodes, although carbon paste electrodes have also been described for this purpose.

The application of polymers functionalized with coordination groups (such as the dithiocarbamate ligand) have been actively investigated. The polymer provides the three-dimensional network along with physical stability, while judicious choice of the ligand controls the selectivity of the sensor. The analyte must be able to permeate the film and bind to polymer bound ligand sites. The formation constant of the complexes K_f is of fundamental importance as it affects the operational performance of the sensor. Large K_f values favor optimum sensitivities and LOD. Ligands with large K_f can be used to effect very selective binding, thus discriminating against possible interferents with smaller formation constants. Large K_f values can also be a disadvantage, as the surface concentration of complexing sites is low, typically 10^{-8} mol/cm^2, saturation of the sites and consequently response saturation may occur at low analyte concentrations if the complexation reaction is too favorable. As the modifier participates in a dynamic chemical reaction prior to electrolysis, the time for this reaction to reach equilibrium should be as short as possible for practical application; unfortunately, this is seldom the case. In general, equilibrium times can be controlled by appropriate choice of electrode size, polymer film thickness, polymer morphology, and K_f.

An elegant example of this type of electrode modification has recently been demonstrated. In this work a PVP polymer containing the ligand 2,2'-bipyridyl was found to be effective at sequestering Fe(II) from solution. The incorporated Fe(II) was then oxidized by linear sweep voltammetry. The problem of response saturation was also addressed. By incorporating an "internal standard" into the polymer film, in this case a ferrocene moiety, which can be determined accurately in situ using slow sweep rate cyclic voltammetry, the exact number of coordination sites on the electrode surface can be determined. Knowing the coordination number of the metal of interest from solution studies, one can assess the response (charge) expected from a saturated film. Sufficient mobility of the polymer-bound coordination sites within the polymer is of paramount importance in order to form stable multidentate complexes. It is recognized that ligand mobility within the film may be restricted by steric or other factors, and as a result multiple coordination to a metal center may not take place. As the formal potential of the metal center is a function of its coordination sphere, this shortcoming may have serious implications for analytical application. For Fe(II) coordination, a tris complex was formed (as expected from solution studies) within the polymer, and consequently no restrictions on ligand mobility was evident. In situations where ligand mobility may be problematic, it has been suggested that thin polymer films may alleviate this matrix effect. In another study it was found that only 0.1–1.0% of the actual analyte incorporated into a chelating film was subsequently detected electrochemically. This suggests mass- and/or charge-transport restriction within the polymer film is detrimental to sensor responses. For this reason such polymers should ideally exhibit fast charge-transport and mass-transport properties to exploit the preconcentration effect fully.

It is fortuitous that selectivity trends for metal–ligand interactions exhibited in solution is maintained in coatings on an electrode surface. With the wide body of information concerning such interactions in solution along with the huge number of ligands available, considerable scope exists for the development of novel electrochemical sensors from these materials.

The potential of ion-exchange materials for electrode modification was recognized early in the development of modified electrodes. Some of the first polymer-modified electrodes described were based on the electrostatic incorporation of charged redox sites into polymers with an opposite charge. The ability to preconcentrate electroactive material in a film close to the electrode surface using the inherent thermodynamic driving force for the ion-exchange reaction resulted in several applications of this process for electroanalysis. The most widely studied materials for ion-exchange preconcentration are protonated PVP; quaternized PVP, which are polycationic; and Nafion, which is an anionic ionomer.

The fundamental behavior of ion-exchange processes that can occur with Nafion and other ion-exchange materials have been investigated extensively. For any ion-exchange material, the extent of preconcentration depends on the ion-exchange selectivity coefficient K_{ex}. Preconcentration occurs when the selectivity for the analyte is greater than that for the competing counterion. For Nafion, ion-exchange selectivity coefficients depend on the charge and structure of the exchangeable ions. In general, selectivity coefficients increase dramatically with increasing charge and hydrophobicity of the ion. The bulk structure of Nafion in electrolyte is thought to be composed of a hydrophilic domain containing $-SO_3^-$ groups and a hydrophobic domain constituting the fluorocarbon framework (see Figure 3.20). It is this structure which results in high selectivity for large hydrophobic cations. For quaternized PVP-based modifiers, the greater the charge on the anion, the more effectively it will be scavenged from solution. As the extent of preconcentration is determined by the molecular properties of the analyte and any competing ions in the sample, considerable matrix effects can be encountered in real analytical situations using these types of materials. Some control over ion-exchange selectivity can be achieved by careful control of electrolyte pH as this affects the ionization of weak acids and bases and consequently their availability for ion exchange. Also the ionic strength of the electrolyte should be kept as low as possible to prevent competition from electrolyte co-ions. However, because of the general nature of ion-exchange processes, ion-exchange preconcentration is not as selective as a coordination-based process.

One of the major disadvantages of ion-exchange preconcentration is the lengthy time intervals required for the reaction to reach equilibrium. The equilibrium time depends on the size, charge, structure, and concentration of analyte; the ionic strength of the electrolyte; geometric dimensions of the electrode; and the thickness of the polymer film. It is usually found that the rate-determining step in reaching equilibrium is the rate of ionic diffusion within the film; for this reason polymers must be designed for maximum mass-transport rates. It has also been observed that ion-exchange isotherms are linear over a small concentration range; therefore, these electrodes may be easily saturated and have a limited response range.

Probably the most popular application of ion-exchange preconcentration is the determination of dopamine. Dopamine is cationic at physiological pH levels and can be selectively incorporated into Nafion. This preconcentration step, along with the exclusion of ascorbic acid, allows voltammetric analysis of dopamine in vivo.

Analytes can also be preconcentrated by reaction with chemically reactive groups at the electrode surface. This approach may result in considerable selectivity enhancement due to the specific nature of the chemical reaction. It is possible that stereospecific reaction may be used to distinguish between optical isomers.

3.6.3.2. Permselectivity

The use of polymeric permselective films on electrode surfaces provides an elegant approach for discriminating the species of interest over coexisting interferences. Permselective membranes provide an in situ separation process directly at the electrode surface, resulting in an increase in selectivity and provide protection of the electrode surface. This is achieved by controlling mass transport through the polymer film in such a way that only the analyte is able to permeate the polymer film. It should be emphasized here that the focus of permselective film design is to optimize the rejection of potential interferents and allow distribution of the analyte into and transport through the polymer film. Obviously permselective films need not preconcentrate analytes; it is, however, advantageous to provide such an effect as this may offset the lowering in sensitivity due to hindered mass transport through the polymer film. Usually the polymeric films employed are electroinactive, but examples of permeation control of conducting polymers have also been demonstrated. Selective discrimination of interferences are based on molecular properties such as charge, size, shape, and polarity.

Cellulose acetate has been a very successful electrode modifier for the exclusion of interferents based on molecular size. The porosity of cellulose acetate can be varied by time-controlled base hydrolysis of the polymer on the electrode. This treatment results in uniform holes in the polymer film through which small analytes can diffuse to the underlying electrode and prevent passage of larger electroactive ions and neutral organic surfactant molecules. The permselective effect of base-hydrolyzed cellulose acetate as an electrode modifier can be seen in Figure 3.21. This material prevents surface passivation due to protein adsorption, producing reproducible responses to uric acid. In the absence of a cellulose acetate, surface passivation occurs quickly and results in a decrease in electrode response. It can also be seen that electrode sensitivity is reduced as a result of diffusion constraints imposed by the film on the analyte.

Polyaniline, polyphenol, and polypyrrole have also been used as permselective membranes. The permselectivity of these materials was found to be dependent on the electropolymerization time and the concentration of the monomer. From the results of a fundamental study, the selective permeabilities of polyphenol and polyaniline can be explained approximately on the basis of the Stokes radii of the

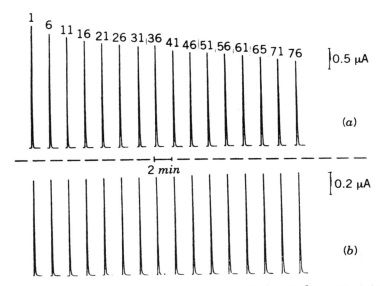

Figure 3.21. Detection peaks for repetitive injections of a 1×10^{-4} mol/dm^3 uric acid solution, containing 200 mg/dL bovine albumin: (*A*) bare electrode, (*B*) electrode coated with cellulose acetate; flow rate 0.6 ml/min; film hydrolyzed for 40 min; [From J. Wang and L. D. Hutchins, *Anal. Chem.* **57**, 1536 (1985).]

permeating ion. However, the degree of selectivity of the films cannot be explained by the differences in Stokes radii alone. It has been proposed that this may be due to a chemical interaction between the permeation species and the modifier.

It was stated earlier that Coulombic association can be used to preconcentrate analytes in polymeric films through ion exchange. The opposite force, coulombic repulsion can be used to prevent interferents from reaching the electrode surface. This is based on the principle that charged polymer films tends to reject ions with the same charge (co-ions). The extent of rejection depends on the net charge on the co-ion and on the polymeric film, degree of swelling and thickness of the polymer film, type, and concentration of electrolyte and electrolyte pH. Unfortunately, a small amount of diffusion of co-ions through charged polymer films is quite a common phenomenon, resulting in nonzero permeabilities. The extent of the process must be assessed to allow application of these materials for analysis.

3.6.3.3. Electrocatalysis

Electrochemical reactions differ from chemical reactions in that the rate constant for the reaction is dependent on the potential of the electrode. Theory dictates that electron transfer be isoenergetic (i.e., the energy of the electron and accepting/donating orbital be equal) so at a potential where isoenergetic conditions are satisfied electron transfer should be in equilibrium. This is the formal potential for the reaction. In reality it is often required to apply potentials in excess of that required by thermodynamics to drive the electrochemical reaction. This excess potential (energy) is known as the *activation overpotential*. In general, one wishes to avoid the application of activation overpotentials. From the perspective of electroanalysis, the application of potential extremes causes an increase in the number of possible interferences. Consequently, reducing operational potentials by even a small amount may considerably reduce interferences. Modified electrodes have been utilized extensively to achieve this. In addition, with redox active modifiers, acceleration of the electrode reaction can be obtained, resulting in considerable increases in the sensitivity and lowered LOD for the electrochemical method. Many materials immobilized on electrode surfaces have been used for the electrocatalysis of electrode reactions. These include metallic and metal oxide microparticles immobilized in polymer matrices, organic redox polymers, and redox polymers containing a coordinated electroactive group. As an example, the electrocatalysis of the Fe(III) reduction at a $[Os(bpy)_2(PVP)_{10}Cl]Cl$-coated electrode is shown in Figure 3.22. This reduction process is ill defined at a bare glassy carbon surface and requires a high potential. By simple modification of the surface, the reduction occurs at much lower potential, is well defined, and is very reproducible.

Modified electrodes for electroanalysis usually accomplish electrocatalysis with the use of surface immobilized redox sites that shuttle electrons to/from reactants via outer-sphere electron transfer. This process is depicted in Figure 3.23. One prerequisite for any application of a modified electrode is that efficient charge transport is taking place throughout the layer. Only in this way can efficient contact be maintained between the solid electrode, which acts as the electron sink, and the analyte solution. This also ensures that the electrode response is a function of the analyte concentration and not of the mobility of the electrons within the film. It is always assumed that electrochemical contact between the underlying solid electrode and the modifying layer is very fast and will not affect the electrochemical processes occurring within the layer. The transport of charge through the polymer is thought to take place via an electron-hopping mechanism. The electron hops from one redox active center to the next. This process is, however, not necessarily the rate-determining step. The polymer is visualized as a porous layer saturated with electrolyte, as may be seen in Figure 3.24. Depending on the nature of the polymer coating, three processes can potentially control the charge-transport process within the layer: polymer movement, electron movement, and ion movement. The properties of the polymer layer will determine which of these three processes will be the slowest and therefore the rate-determining one. In many polymers, especially in conduction polymers, electron hopping is very fast and ion

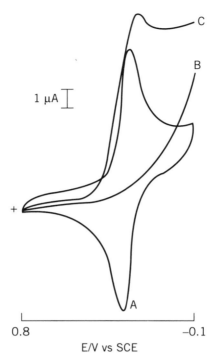

Figure 3.22. (*A*) Cyclic voltammogram of $[Os(bpy)_2(PVP)_{10}Cl]Cl$ in 0.1 mol/dm^3 H$_2$SO$_4$, scan rate 10 mV/s. Reduction of 8×10^{-4} mol/dm^3 Fe(III) at (*B*) a bare glassy carbon electrode and (*C*) a glassy carbon electrode modified with the osmium metallopolymer.

movement or chain movement are rate determining. Ion movement is important as, on oxidation of the redox active center, charge-compensating counterions have to be brought into the film.

When the conditions for efficient charge transport have been optimized, the mediation process can be considered. A schematic diagram of a polymer-modified electrode is shown in Figure 3.23. In this diagram the "A" stands for the polymer-bound redox center, and the analyte as "Y." This centre can be electrostatically or covalently bound to the polymer backbone. This redox active center will control the electrochemical signal obtained, and reaction of the analyte with the underlying electrode should not occur if we wish to obtain optimum benefit from the coating. This mediation process can be described by the following reactions:

$$A \rightarrow B + e^- \tag{3.12}$$

$$B + Y \rightarrow A + Z \tag{3.13}$$

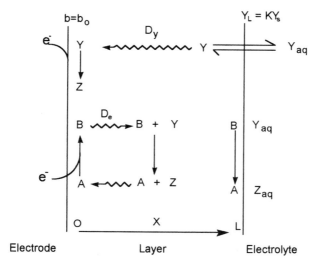

Figure 3.23. Schematic diagram of a modified electrode containing the redox couple A/B involved in a mediated reduction of Y to Z.

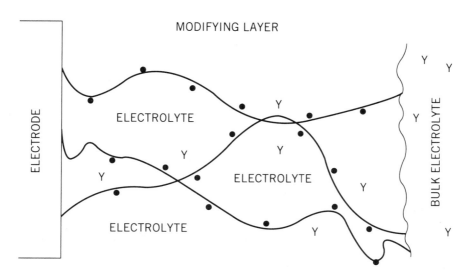

Figure 3.24. Schematic diagram of a swollen polymer film at a electrode surface.

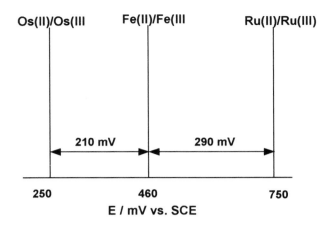

Figure 3.25. Comparison of the redox potentials of the redox catalysts [Os(bpy)$_2$(PVP)$_{10}$Cl]Cl and [Ru(bpy)$_2$(PVP)$_{10}$Cl]Cl with that of the Fe(II)/Fe(III) redox couple.

In this example the mediation process is based on the oxidation of the surface-bound redox couple, but for an reduction process, the same approach can be taken. Whether not reaction (3.13) is in fact taking place is determined by the formal potential of the A/B and Y/Z redox couples. A careful selection of the $E^{0'}$ of the electrocatalytic center is therefore of great importance. For any mediated cross-exchange reaction, the Gibbs free energy for the reaction must be negative. This imposes certain limitations on the reactions that can occur at redox polymer-modified electrodes. The implication is that for a reduction reaction, the formal potential of the analyte must be positive of the $E^{0'}$ of the electrocatalyst; and for mediated oxidation reactions, the formal potential of the analyte must be negative of the $E^{0'}$ of the electrocatalyst. This is visualized in Figure 3.25. The closer the redox potential of the electrocatalyst is to the redox potential of the analyte the more selective the sensor is. This, however, also reduces the driving force of the reaction and consequently the sensitivity. The choice of redox catalyst must therefore be based on both the concentration expected for the analyte and the types of interference expected. Figure 3.25 is of relevance for the speciation of Fe(II) and Fe(III). It shows the formal potentials of two electrode modifiers, [Os(bpy)$_2$(PVP)$_{10}$Cl]Cl and its ruthenium analog (for structures, see Figure 3.17). For the ruthenium polymer, a redox potential of 750 mV versus SCE is observed, while for the osmium analog, the redox potential is 250 mV versus SCE. In general, the oxidation of compounds having a less positive redox potential than that of a modifying layer can be mediated by this material, but not their reduction. In a similar way for a compound having a more positive redox potential than that of the coating, only the reduction process can be mediated. It can be seen from Figure 3.25 that the redox potentials of the osmium and ruthenium polymers envelope the formal potential of the Fe(II)/Fe(III) redox couple, which is 460 mV versus SCE. From the positioning of the redox couples, it can be concluded that the ruthenium coating will catalyze the oxidation of Fe(II), with a driving force of 290 mV, while there is a driving force of 210 mV for the reduction of Fe(III) by the osmium electrode. As a result, when an electrode modified with the ruthenium polymer is held at a potential where the ruthenium is in the Ru(III) state, the oxidation of Fe(II) is catalyzed as in reactions (3.12) and (3.13). On the other hand, if an electrode containing the osmium polymer is held at 250 mV, where the osmium centers are in the osmium(II) state, the reduction of Fe(III) is catalyzed and the reverse reactions will not occur. As a result, a dual-electrode system can be developed with these polymers that will simultaneously detect Fe(II) and Fe(III), something that is not possible with conventional electrodes. The Fe(II)/Fe(III) speciation sensor is constructed from two modified glassy carbon electrodes placed in a parallel arrangement in a thin-layer electrochemical cell. One electrode is modified with $[Ru(bpy)_2(PVP)_{10}Cl]Cl$ and the other one, with its osmium analog. At an applied potential of 0.8 V versus SCE, the ruthenium electrode can mediate the oxidation of Fe(II) selectively, while at the osmium electrode the Fe(III) reduction is mediated at a potential of 0.250 V. This allows the simultaneous detection of the two species in a single detection system. The modification of the electrodes not only greatly improved the redox behavior of the Fe(II)/Fe(III) redox couple by lowering the operation potentials but also improves the electrode response by a factor of 5. For both electrodes mediation occurs throughout the polymer film and results in a LOD of 5×10^{-8} mol/dm^3 and a linear range from 1×10^{-7} to 1×10^{-2} mol/dm^3. Typical traces are shown in Figure 3.26.

The preceding example shows the potential of a modified electrode if the redox potential of the modifier is matched with that of the analyte. The applicability of the modified electrode is therefore determined mainly by the redox potential of the redox couple. The type of redox active centers can vary from metal complexes such as ruthenium polypyridyl complexes and ferrocene to organic redox couples such as methylviologen to enzymes and antibodies. The polymer backbone can be both conducting, as for polypyrrole and polyaniline; or nonconducting, as for polyvinylpyridine or polyvinylimidazole. If a coating is only applied to prevent fouling at the solid electrode or to preconcentrate the analyte before the measurement, the electrochemical reaction does, of course, still take place at the underlying electrode, and the mechanism as described below is not relevant and the currents observed can be explained by conventional analysis.

At present polymer-coated electrodes are most promising in the development of new, more specific sensors. Not only does the use of polymers give a wide range of synthetic methods, but most importantly, with polymer coatings one can, in principle, obtain a three-dimensional sensor. That is, the whole polymer layer is active in the measuring process and in this manner the sensor active surface is much larger than for a traditional electrode. As this will greatly increase the sen-

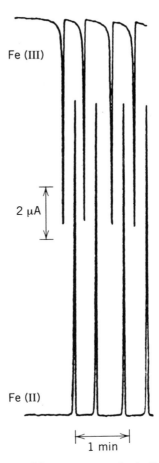

Figure 3.26. Typical recorder traces of detector responses for the flow injection analysis detection of Fe(II) and Fe(III) using the redox polymers $[Ru(bpy)_2(PVP)_{10}Cl]Cl$ and $[Os(bpy)_2(PVP)_{10}Cl]Cl$, respectively. Fe(II) and Fe(III) are 1.0×10^{-3} mol/dm³; electrolyte 0.2 mol/dm³ Na_2SO_4/pH = 1.0; flow rate 1.0 mL/min, surface coverage for each polymer 1.0×10^{-9} mol/cm². [From A. P. Doherty, R. J. Forster, M. R. Smyth, and J. G. Vos, *Anal. Chem.* **64**, 572 (1992).]

sitivity of the sensor, a through-layer measurement offers a great advantage. Whether the whole layer is involved in the measuring process is, however, dependent not only on the structure of the polymer but also on the measurement conditions. So the most efficient sensor is obtained when a large part of the layer participates in the reduction or oxidation of the substrate, Y. Whether this will take place is for a large part dependent on the relative importance of electron and substrate diffusion and of the rate of reaction (3.13). These first two parameters will be strongly dependent on the structure of the mediating film. The reaction rate depends on the driving force of reaction (3.13). Another factor is the concentration of the redox active sites. In order to obtain an efficient turnover, the redox site concentration should be higher than that of the substrate. This is in general quite easily achieved as the coatings are very compact. This has the result that very high local concentrations of mediating sites can be obtained. An example of a typical coating is given in Figure 3.17. In this polymer one out of every 10 pyridine groups is coordinated to a redox active osmium center. Because of the size of the pendant group the highest loading that can be obtained is 1.5. Density measurements have shown that the concentration of the redox centers in the dry materials is of the order of 1 mol/dm^3, and therefore much higher than the average solution concentration of analytes. It is, however, important to realize that a very high redox site loading does not necessarily give the most favorable mediating conditions. It has been shown that very high ($>$1 mol/dm^3) redox site concentrations inhibit the diffusion of counterions and substrate in these films and intermediate loadings can therefore give better results. So it is important that ions and substrate can freely enter the coating. In Figure 3.24 it is assumed that large parts of the coating can contain solvent and electrolyte. This is, however, not always the case. Very heavily cross-linked polymers, or a polymer in contact with a nonsolvent for that polymer, will often not allow significant penetration of the electrolyte, and the electrochemical process can take place only at the surface of the modified electrode as is shown in Figure 3.23.

3.7 AMPEROMETRIC BIOSENSORS

3.7.1. Introduction

Amperometric biosensors are analytical devices in which a biological material is used as a biological catalyst in combination with an electrical transducer. A biosensor responds to an analyte in a sample and interprets its concentration as an electrical signal via a biological recognition system and the electrochemical transducer. Electrochemical biosensors have been under development for 30 years, and over this time a wide variety of sensors has been developed. The range of biological catalysts include enzymes, antibodies, chemoreceptors, cell organells, and cellular tissue. The overriding theme of biosensors is the ability to perform selective biological recognition of the target analyte in a complex sample matrix and couple this to the sensitivity of electrochemical detection. Undoubtedly, the most successful and best understood biosensors are those using enzymes as their mode of molecular recognition.

The magnitude of the response of amperometric biosensors depends on a number of factors, including the kinetics of the enzymatic reaction, the construction of the enzyme electrode, and the mode of operation of the electrode. The response from the electrode can either be diffusionally controlled or kinetically controlled. With kinetically controlled enzyme electrodes, the enzyme loading is sufficiently low that the response depends on the concentration of enzyme and the kinetics of the enzymatic reaction. Such behavior has limited analytical utility as response saturation occurs at low substrate concentration. The diffusionally controlled electrode possesses very high enzyme loadings such that the current is independent of small changes in enzyme concentration; consequently, the current response is a function of analyte concentration and diffusion.

The construction of the enzyme electrode affects the response considerably. The effect of enzyme concentration, enzyme layer thickness, and the semipermeable membranes used have been investigated in detail, along with extrinsic parameters such as pH, temperature, analyte diffusion, and concentration. In general, forced transport of analyte (e.g., stirring of test solution or rotating electrode) using high concentrations of enzyme with thin hydrophilic membranes are recommended.

Enzyme electrodes can be operated in several measurement modes: dynamic, steady-state, potential-step, and flow-injection mode. The steady-state mode allows reaction equilibrium to be reached before the analytical signal is obtained, whereas in dynamic measurement the signal is obtained quickly as a predetermined timepoint after introduction of the sample. Both potential step and FIA measurements are transient responses due to the transient nature of the techniques. Mathematical models describing sensor responses have been described, but because of the complexity and multiplicity of designs, a general theory for all arrangements cannot be formulated.

3.7.2. Enzyme Electrodes Based on Measurement of Oxygen or Hydrogen Peroxide

The concept of the enzyme electrode was first proposed in 1962 as an extension of their successful amperometric oxygen electrode. This concept involved placement of an enzyme in close proximity to an electrode surface, where the enzyme is able to catalyze a reaction involving the analyte with the consumption of an electroactive reactant (O_2) and/or the production of an electroactive product (H_2O_2). The production or depletion process is monitored amperometrically and gives a direct measure of the analyte concentration. The enzyme electrode has undergone continuous development since its introduction, and it is convenient to discuss progress in terms of "generations." In recent years many enzymes have become commercially available through developments in genetic engineering and improved enzyme purification procedures; in addition, the advances in technology for immobilization and increased availability of electrical transducers have resulted in the rapid development of enzyme biosensors.

Based on this concept the first enzyme electrode was produced in 1967. The electrode was constructed from polyacrylamide gel impregnated with glucose oxidase placed in intimate contact with the Clark oxygen electrode. The following enzyme reaction occurs at the electrode surface:

$$Glucose + FAD + H_2O \rightarrow gluconic\ acid + FADH_2 \qquad (3.14)$$

$$FADH_2 + O_2 \rightarrow FAD + H_2O_2 \qquad (3.15)$$

where FAD = flavin adenine dinucleotide, the prosthetic group of glucose oxidase. Oxygen is the natural electron acceptor for the oxidation of the flavoprotein glucose oxidase. Its small size permits entry into the active site of the enzyme, where it oxidizes the reduced form of the prosthetic group ($FADH_2$). Glucose from the sample diffuses into the polyacrylamide gel, where it undergoes the enzyme-catalyzed oxidation reaction with O_2. With the O_2 sensor held at the reduction potential of O_2 (-600 mV vs. Ag/AgCl), the decrease in the steady-state response to O_2 diffusing into the gel from the sample is a direct measure of glucose concentration. Alternatively, the production of H_2O_2 can be monitored using an electrode held at 600 mV versus Ag/AgCl. As O_2 is the electron-transfer mediator, this type of electrode is classified as a first-generation enzyme electrode. Many examples of this generation of enzyme electrode can be found in the literature using various procedures to immobilize the enzyme.

The simplest immobilizing procedure involves placing a paste of the enzyme directly on the electrode surface followed by a protective layer of dialysis membrane. More sophisticated devices involve adsorption of the enzyme onto the electrode surface. This technique is simple, requires no reagents, and causes minimal disruption to the enzyme. This approach, however, is more susceptible to changes in pH, temperature, ionic strength, and support and also requires considerable optimisation. This method is not widely applicable because of leaching of the biocomponent and the relatively short lifetimes of the enzyme. Adsorption is usually carried out by evaporation of a buffered solution of the enzyme on the electrode surface at 4°C.

It has been shown that platinization of platinum, to form platinum black, provides a good support for the adsorption of enzymes due to the large surface area created. The enzyme may be crosslinked on the electrode surface to form a thin film using glutaric dialdehyde and bovine serum albumin. This procedure is thought to increase the rigidity of the structure of the enzyme, thus maintaining the native conformation of the protein and rendering unfolding less probable, hence increasing operational stability. In another study, glucose oxidase was adsorbed onto a carbon paste electrode by solvent evaporation. This was then treated with a collodion–ethanol solution to produce a thin film of nitrocellulose to retain the enzyme on the surface.

Covalent attachment of enzymes to electrodes is technically more difficult than adsorption, but it provides a more stable biocomponent. It is more widely applicable and is usually the method of choice if long operational performance is required. Covalent attachment normally involves three steps: activation of the solid supporting electrode, enzyme coupling, and finally removal of unbound enzyme. Support activation is normally accomplished by oxidation of the electrode surface followed by selective reduction with $LiAlH_4$. The surface is then treated with chemicals such as silanes and cyanuric chlorides to introduce surface functionalities. The enzyme is then attached directly to the functionalized support or via a difunctional coupling agent. Oxidized graphite surfaces can also be activated with soluble carbodiimines under acidic conditions. Covalent coupling can then be accomplished through an amine linkage between the carbonyl on the electrode and an amine on the enzyme. The coupling reaction should involve functional groups in the enzyme that are not involved with the catalytic reaction. Amino, thiol, and hydroxyl groups on the enzyme may participate in the coupling reaction. The amino acid residues cysteine, lysine, tyrosine, and histidine are considered the most reactive because of the large number of coupling reactions they can undergo. This implies a considerable risk to the enzyme during immobilization, but partial loss of activity may be the price to pay for long operational performance.

The most popular technique for the immobilization of enzymes at electrodes involves gel or polymer entrapment. As mentioned earlier, this may simply involve the physical trapping of an enzyme paste at an electrode surface with materials such as cellulose acetate. Alternatively, the enzyme may be physically or chemically immobilized in the gel/polymer layer. With these materials the substrate of the enzymatic reaction diffuses from solution into the membrane, and on reaching the enzyme, the product will be formed, after which it will diffuse to the electrode surface and be detected. Such membranes should have good adhesion to the electrode, and should be thin, hydrophilic, and porous. It should be remembered that an enzyme electrode is a dynamic device unlike solid-state electrodes. Typical polymers include poly(vinyl alcohol), polyurethane, cellophane, polyacrylamide, agarose, and polypyrrole. The use of gels/polymers allows mild conditions to be used with a wide variety of gels and polymers. This technique, however, suffers from several disadvantages; many experimental factors must be optimized and controlled, deactivation of the enzyme may occur due to reaction with radicals produced during the polymerization process, and the macroscopic nature of the film may result in diffusional constraints that can increase response times considerably.

The coimmobilization of glucose oxidase with polypyrrole by electropolymerization onto a platinum electrode at pH levels above the pK_a of pyrrole has been demonstrated. This results in an anionically charged film that appears to electrostatically entrap the enzyme in the film as a counterion. Polypyrrole is a conducting polymer, but in situations where O_2 is the electron mediator, H_2O_2 generated by the enzyme reaction destroys the intrinsic conductivity of the polymer. It has been shown that the H_2O_2 is oxidized at the underlying electrode; therefore, polymer conductivity is not a prerequisite for successful sensor operation. Polymer conductivity is, however, advantageous for the formation of thick polymer films to allow large enzyme loadings.

Enzyme electrodes based on the electropolymerization of phenolic compounds to produce films with permselective and surface protecting properties have been investigated. The electropolymerization of polyaniline in the presence of glucose oxidase has also shown to be effective in entrapping glucose oxidase and providing some protection from larger interferences. Examples of enzymes immobilized on nylon meshes have also been demonstrated. Crosslinking of gels and polymers has also been carried out to improve their operational stability.

Polymeric films have been used for purposes other than enzyme immobilization, for example, to present a diffusional barrier between the sensing device and the solution. Such membranes inhibit the diffusional mass transport of the analyte such that the overall catalytic reaction rate is limited by the diffusion process. This principle can be used to control the linear response range of the electrode. In situations where the rate constant for the enzymatic reaction is low, response saturation occurs at low substrate concentration. A passive, diffusion-controlling membrane can be used to avoid response saturation. Other membranes can act actively if the membrane contains functionalities such as ion-exchange sites or complexing groups. Membranes can also be used to prevent precipitation of insoluble material on the electrode surface and also passivation reactions.

Many other electrodes based on the monitoring of O_2 consumption or H_2O_2 production have been described; these include choline oxidase, acetylcholine esterase, cholesterol oxidase, and urease.

The combination of several enzyme systems can be used to produce a selective sensor. Several enzymes can be coimmobilized in a single layer or immobilized separately in several contiguous layers. In the latter case these are generally separated by semipermeable membranes to avoid migration of sample components (which may interfere in subsequent steps) or the biologically active material. Examples of these types of devices include the sucrose sensor and the α-amalose sensor, the operation principle of which can be seen in Figure 3.27.

Multiple-enzyme systems appear quite useful and open up many possibilities for sensor development. It has been pointed out, however, that the deliberate linking of several enzyme reactions becomes uneconomical after about five enzymes and also increases the possibility of side reactions that are difficult to observe and control. The tortuous reaction sequence can also lead to very long response times.

The use of Langmuir–Blodgett (LB) membranes for enzyme immobilization has also been described. Immobilization through hydrophobic or electrostatic interactions in the film generally results in a significant loss in enzyme activity; therefore, LB systems incorporating functional groups that can be used for covalent binding of the enzyme have been introduced. Such electrodes have been constructed from stearylamine to which glucose oxidase has been bound via glu-

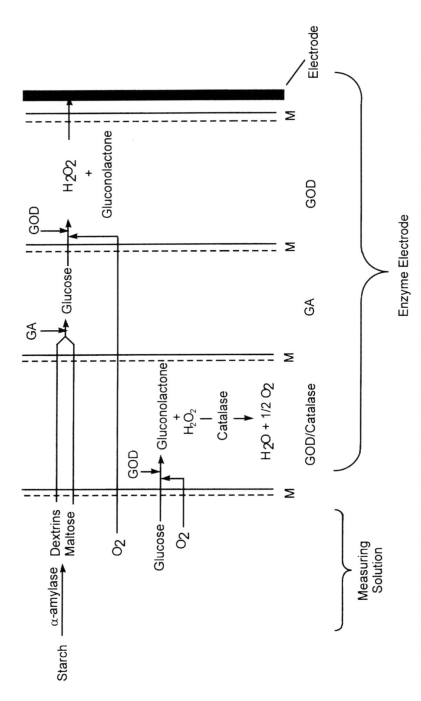

Figure 3.27. Principle of α-amylase and starch determinations in glucose containing samples with a bienzyme electrode covered with a glucose oxidase–catalase antiinterference layer (GA = glucoamylase; M = membrane). [From J. E. Frew and H. A. O. Hill, *Anal. Chem.* **59**, 933A (1987).]

taraldehyde. Other examples include the coimmobilisation of the enzyme in the LB monolayer.

3.7.3. Amperometric Biosensors Based on Incorporated Electron Mediators

Although the O_2/H_2O_2 monitoring schemes appear convenient for sensor design, several limitations limit their use in real samples. Monitoring of H_2O_2 production at 600 mV versus Ag/AgCl introduces the possibility of interferences from easily oxidized substances such as ascorbic acid and uric acid; these compounds are frequently encountered in real biological and clinical samples. H_2O_2 can be consumed by impurities in the enzyme preparation, and this can give rise to erroneous results. H_2O_2 is a strong oxidant and at high local concentrations may deactivate certain enzymes. The electrocatalytic properties of platinum black allow for the oxidation of saccharides and the affinity of platinum black for proteins makes it very susceptible to fouling.

The electron-transfer process involving oxygen is kinetically slow and cumbersome and is both diffusionally and kinetically rate-limiting. In situations where the concentration of O_2 is low (such as in whole blood), the concentration of unbound O_2 may be lower than that of the analyte, which results in a stoichiometric limitation of the enzyme reaction by oxygen. This may result in response saturation at low analyte concentrations. One approach used to try to overcome the diffusional limitation of oxygen has been the two-dimensional electrode, where O_2 diffuses from two directions while analyte diffuses from one direction only.

Another, more successful approach used to overcome both the diffusional and kinetic limitations of O_2 mediation is the use of an artificial electron-transfer mediator to replace the O_2 as the electron acceptor for the reduced form of the enzyme ($FADH_2$). Enzyme electrodes incorporating such mediators constitute the second generation of enzyme electrodes. *Mediators* are low-molecular-weight redox couples that shuttle electrons from the redox center of the enzyme to the surface of the indicator electrode. During the catalytic cycle the mediator, M_{ox}, first reacts with the reduced form of the enzyme and then diffuses to the indicator electrode, where it undergoes rapid oxidation. This process is represented as follows:

$$\text{Glucose} + \text{FAD} \rightarrow \text{gluconic acid} + \text{FADH}_2 \qquad (3.16)$$

$$\text{FADH}_2 + \text{M}_{ox} \rightarrow \text{FAD} + \text{M}_{red} + 2\text{H}^+ \qquad (3.17)$$

At the electrode, the reaction

$$\text{M}_{red} \rightarrow \text{M}_{ox} + e^- \qquad (3.18)$$

occurs if the reduced form of the mediator does not react with O_2. Good mediators should have the following characteristics:

1. Exhibit reversible kinetics
2. React rapidly with the reduced form of the enzyme
3. Have a low oxidation potential and be pH independent
4. Be stable in both redox forms
5. Be easily retained at the electrode
6. Be unreactive toward oxygen

The first artificial redox mediator used was quinone, which is reduced to hydroquinone and is subsequently reoxidized at the electrode. The lower operational potential of the indicator electrode, 400 mV versus Ag/AgCl, was found to reduce the effect of several interferences. Several other mediators have since been used, including thionine, Methylene Blue, and 1,4-naphthoquinoline.

Recently a new generation of mediators based on ferrocene [see Eq. (3.3)] and its derivatives have been introduced. Ferrocene has a well-behaved one-electron redox couple with a low formal potential of 165 mV versus SCE. Substitution of one or both of the cyclopentadiene moieties results in a series of compounds with different charges, solubilities, and formal potentials.

The physical properties of mediators, to a great extent, determine their usefulness and mode of application. It has been pointed out that positively charged ferrocenes are much better mediators than are negatively charged ones. The solubility of the mediator is also important, as this can determine the exact mode of operation. Mediators can be used as a free solution species in the sample, from where it can diffuse into the enzyme layer and to the electrode surface. This mode of operation was demonstrated with theophiline oxidase and the cofactor ferric cytochrome c physically entrapped in Nafion on a platinum electrode, where ferricyanide acted as the mediator. By far, the most common mode of operation is the immobilization of the mediator on the electrode surface by adsorption; this is achieved by evaporation of a solution containing the mediator on the electrode surface. Many mediators are hydrophobic; therefore, they must be applied with organic solvents and consequently the mediator must be immobilized prior to enzyme immobilization to prevent denaturation. The mediator 1,1-dimethylferrocene was adsorbed onto carbon from a toluene solution; the enzyme glucose oxidase was then covalently attached to the electrode surface via a carbodiimide linkage. This system was found to be insensitive to O_2 and had fast response times with a linear range over the concentration range of diabetic blood samples and could be applied to whole blood. Therefore, the sensor could be used without manipulation of the sample. This sensor design has been commercialized successfully.

The use of Meldola Blue as a mediator has also been reported. This mediator adsorbs spontaneously on graphite and oxidises reduced NADH at 0.0 mV versus SCE. The method and order of mediator adsorption on the electrode surface are important. This mediator does not depend on diffusion for charge transfer but by intermolecular charge transfer (electron hopping); therefore, the mediator molecules must be in close proximity with each other.

Ferrocene can also be used to chemically modify other molecules such as polymers to expand their range of application. The glucose oxidase/polypyrrole electrode developed by Foulds and Lowe can be adapted for mediated electron transfer by using monomeric pyrrole functionalized with ferrocene derivatives. This allows the electropolymerized polymer to act as an electron mediator. A disadvantage with this approach is the long response times found due to the limited mobility of the polymer-bound mediator compared to free ferrocene. Recently a "redox wire"-type enzyme electrode was demonstrated that used a substituted version of the earlier mentioned poly-4-vinylpyridine complex of $[Os(bpy)_2Cl_2]$. This polycationic redox polymer forms electrostatic complexes with polyanionic glucose oxidase in a manner mimicking the natural attraction of redox proteins for enzymes. The polymer mediates electron transfer by an "electron hopping" process between fixed redox sites on the electrode surface. In an extension of this work the enzyme was chemically bound to the PVP backbone via N-hydroxysuccinimide, which forms amine bonds with lysine groups on the enzyme surface. However, competition between the aminolysis of the N-hydroxysuccinimide and their spontaneous hydrolysis resulted in significant interfilm variability. This situation was improved by the use of a commercially available diepoxy that linked amino groups on the quaternized PVP with amino groups on the enzyme.

Binding of an enzyme covalently to a polymeric film on an electrode surface requires a binding reaction that can occur at physiological pH under aqueous conditions and results in a polymer that promotes the stability of the enzyme. It is also desirable that the polymer exhibit fast diffusion of substrate and products and rapid charge-transport properties with fast "electrical communication" with the active site of the enzyme.

It has been pointed out that "enzymes hate conventional electrodes and vice versa." The redox centre in the enzyme is surrounded by a protein coat, which makes it difficult for direct electron transfer to or from electrodes. The concept of direct electron transfer between the active site of the enzyme and the electrode surface has been a long-term objective of electrochemists. Direct electron transfer is normally impossible because of the distance between the enzyme active center and the electrode as the rate of electron transfer between molecules and electrodes decreases exponentially with increased distance between them. For proteins, the electron-transfer rate drops by a factor of 10^4 when the distance between the electron donor and acceptor increases from 8 to 10 Å. It is believed that for direct electron transfer to occur to a redox protein, posturing of the macromolecule must take place to orientate the molecule into a position where electron transfer is possible. This often leads to adsorption of the protein molecule. Electrodes in which direct electron transfer between the enzyme and electrode constitute the third-generation enzyme electrodes.

One of the earliest reported examples of direct electron transport was the of cytochrome c at a gold electrode in the presence of 4,4'-bipyridyl. The haem group of the cytochrome c undergoes a rapid almost reversible redox process at the modified gold electrode. It is postulated that the 4,4'-bipyridyl molecules adsorb onto the gold surface and provides an appropriate interfacial environment for the reversible adsorption of cytochrome c. The adsorption process is believed to involve hydrogen bonding between the surface pyridine and the lysine groups around the exposed haem edge of the protein, mimicking the analogous physiological reaction. This process provides the correct enzyme orientation for rapid charge transfer, and also the considerable binding energy provides a source of activation energy for the electron-transfer step.

Conducting organic salts (organic metals) are particularly good electrode materials for the direct electron transfer to and from enzymes, particularly the NAD^+ cofactor system. Over 250 enzymes use NAD^+ as a cofactor; therefore, there is wide scope for the application of conducting salt electrodes. These materials typically consist of salts comprising two planar organic molecules with extended π-electron systems formed in segregated stacks. Under suitable conditions, overlap of the π-orbitals can lead to delocalization of charge through the stack to give rise to electrical conductivity. This partial transfer occurs from the donor to the acceptor compound. These donor–acceptor complexes are metallic at room temperature. A typical example of a donor–acceptor complex is N-methylphenazinium/tetracyanoquinodimethanine. Conducting salts have also been used with a range of flavoproteins such as glucose oxidase, xanthine oxidase, and choline oxidase. Conducting salt electrodes exhibit advantages such as low operational potentials, very facile electron-transfer kinetics, fast response times, and high stability. The sensing surfaces can be renewed by potential sweeping out of the stable region, which causes dissolution of the organic salt at the surface.

This type of electrode can be made by drop coating a mixture of the conducting salt with polyvinylchloride (PVC) in tetrahydrofuran onto conventional electrodes followed by immobilization of the enzyme by adsorption or entrapped in a polymeric membrane. Direct electron transfer from enzymes to electrodes has been reported by the direct immobilization of ferrocene in the interior of glucose oxidase.

3.8.   RECENT ADVANCES IN AMPEROMETRIC METHODS

As dimensions and timescales of electrochemical experiments shrink, the stochastic nature of individual electrochemical and molecular events will become evident. This has already been studied using a scanning electrochemical microscope (SECM) with a tip of nanometer dimensions where the electrochemical cell has a feature that allows the molecules to be trapped near the tip.

Interesting aspects of the recent literature (see, e.g., the biennial reviews in the journal *Analytical Chemistry*) have included the use of a wall-jet mercury thin-film electrode for square-wave anodic stripping voltammetry that was found to achieve lower limits of detection in comparison to a stationary electrode used under the same voltammetric conditions. Interdigital arrays of microelectrode bands have been used with a reversible electrochemical couple to allow a steady-state current to be obtained similar to that mentioned in Section 3.3.1. A 0.5-fmol detection limit for dopamine at a carbon interdigital array microelectrode detector linked to a HPLC microbore system was obtained as a result of low background noise and redox cycling. Anodic stripping voltammetry has been used as a method of detection for capillary flow-injection analysis to determine trace heavy metals in tears. Expert systems have also been applied to help design experiments for speciating selected oxidation states of particular metals.

Another technique that has become popular is that using screen-printed electrodes. These can be made easily using carbon powder and cellulose acetate in a suitable solvent (such as acetone/butanone mixture) along with standard screen-printing materials. Modification of the ink using cobalt phthalocyanine has been employed to make electrodes more specific for ascorbic acid detection.

Thin, small devices can be mass-produced by screen-printing for the analysis of glucose and for the determination of 3-hydroxybutrate. These devices represent more closely the inroads that electrochemistry have made in the development of sensors and are exemplified by the ExacTech glucose pen, a device that allows diabetics to monitor their own glucose levels easily.

Electrochemistry has also been hyphenated with mass spectroscopy (MS), in the form of differential electrochemical mass spectroscopy and electrospray mass spectroscopy. In this way volatile reagents and products from an electrochemical cell are introduced into MS for characterisation.

One of the more surprising modern techniques has been the incorporation of plant tissue into porous reticulated electrodes and carbon paste (a paste made from finely powdered carbon and a mineral oil). Materials such as banana, eggplant, and horseradish root have been used. The rationale is that these tissues contain enzymes in their natural environments and thus that these enzymes stand a greater chance of acting more efficiently and surviving longer than do prepurified enzymes. The use of enzymes makes electrodes selectively responsive toward particular analytes. For example, horseradish root has been used in an electrode to determine butanone peroxide in drinking water. In keeping with advances in microlithography, not only are there reports of microscale three-electrode cells, being fashioned on chips, but also integrated potentiostats. It is these latter advances that should allow the use of electrochemistry to blossom in the area of sensors and be added to the wide array of transducers based on techniques such as fiber-optic, piezoelectric, and other methods.

One principal aim of biosensor research is the development of implantable sensors for continuous monitoring of critical substrates such as glucose, such measurements are desirable for the precise control of conditions such as diabetes. A number of key requirements for such devices include size, biocompatibility, lifetime and nontoxicity. Recently significant advances have made in this area by bringing together many of the concepts developed since the inception of modified electrodes. One key advance is the direct "wiring" of enzymes to the osmium-containing redox polymers developed independently by Heller and Vos. This allows electrical contact between the enzyme redox center and the electrode via the mediating centers of the redox polymer. This approach overcomes problems with potential leaching of toxic mediator in vivo as these materials have been shown to be highly stable both physically and chemically. Peroxidase-based prelayers have been used with these electrodes to eliminate electrooxidizable interferences such as ascorbate and urate. Nafion layers have also been shown to extend sensor linear ranges by acting as diffusion barriers. These electrodes may be miniaturized to less than 10-mm dimensions and encapsulated in biocompatible materials such as crosslinked poly(ethylene oxide) for implantation and continuous in vivo monitoring with the total sensor dimensions less than 3 mm in diameter. Although problems with sensor lifetimes in vivo have yet to be overcome, this work holds promise for implantable devices and the development of a wide range of mediated enzyme electrodes.

Another area in amperometric sensor development, which has seen rapid development recently, is amperometric immunosensors. Biotechnological advances in immunochemistry have allowed the production of highly specific monoclonal antibodies for binding to analyte antigens, including enzyme-linked immunosorbent assays (ELISA). In the electrochemical version of this highly successful technology, antibodies for the substrate are immobilized onto an electrode surface (e.g., carbon felt) onto which the sample is applied and incubated to allow binding of the analyte (analyte = antigen). Following this, a washing cycle removes the sample matrix and subsequent binding of a second antibody for the analyte is effected. The second antibody incorporates an immobilized enzyme such as alkaline phosphatase; the overall activity of the bound enzyme is a direct function of the surface concentration of antigen (analyte) and hence solution concentration of analyte. Subsequent washing removes excess enzyme-labeled antibody, and the electrode is incubated with an enzyme substrate, such as aminophenyl phosphate. The product of the enzyme–substrate reaction, p-aminophenol, is then detected amperometrically. Recently, one-step separation-free and regenerable amperometric immunosensors have been developed. Various enzymes may be used (e.g., peroxidase) to effect detection via H_2O_2, while the use of soluble electron-transfer mediators such as hydroquinone or potassium ferrocyanide have also been used to effect electrocatalytic amplification of the sensor response. Coimmobilization of the mediator, butyl ferrocene, with the antibody on the electrode surface has been demonstrated to effect direct electron transfer through the immunoglobulin layer. Recently, the use of osmium-containing redox polymers as nondiffusional redox mediators has been proposed in a similar fashion to that developed by Heller for enzyme electrodes to "wire" the analyte antigen (an enzyme) redox center to the electrode surface. By implication, the direct "wiring" of the enzyme of ELISA systems is possible. The coupling of the unique selectivity of immunochemistry and the sensitivity of electrochemistry is producing sensor with limits of detection at the picomolar level; in addition, immunosensors may be applied for the analysis of small molecules such as pesticides, to macromolecules such as enzymes, and to structures such as viruses and bacteria. Future developments in this area will undoubtedly lead to unique, rapid, and reliable sensor devices for a wide range of analytes, not just on the molecular scale.

3.9.   CONCLUDING REMARKS

The aim of a sensing device is to be small, simple, and capable of yielding a value (within specified tolerances) for the concentration of an analyte in a potentially complex matrix. It should be a real-time device for "instant results." The data acquisition aspects for such a sensor have long been overcome, and it rests with researchers to develop suitable transducers. Many of the instrumental methods mentioned in Section 3.3 fulfil the analytical criteria for a sensor; however, they are complex and need substantial investment in equipment. The ideal sensor should present a minimum amount of complexity and produce a result that may not be as accurate as an "analytical instrument," but well within acceptable limits. Until recently, amperometric analysis has been greatly restricted to the use of analytical instrumentation. The recent arrival of chemically and biologically modified electrodes, along with the use of microelectrodes and screen-printed electrode designs, should prove an important catalyst in the development of specific transducer layers. With the introduction of the modified electrode, the possibilities for the design of specific amperometric sensors have increased dramatically, and it is in this area that most advances in modern sensing devices can be expected in the near future.

SUGGESTED FURTHER READING

Bard, A. J., and Faulkner, L. R., *Electrochemical Methods,* Wiley, New York, 1980.

Bartlett, P. N., in A. P. F. Turner, I. Karube and G. S. Wilson, eds., *Biosensors, Fundamentals and Applications,* Oxford University Press, Chap. 13, p. 211, 1987.

Bond, A. M., *Modern Polarographic Methods in Analytical Chemistry,* Marcel Dekker, New York, 1980.

Fleischmann, M., Pons, S., Rolison, D. R., and Schmidt, P. P., eds., *Ultramicroelectrodes,* Datatech, Morganton, 1987.

Hillman, A. R., in *Electrochemical Science and Technology of Polymers,* R. G. Linford, ed., Elsevier, London, 1986, Chapters 5 and 6.

Kissinger, P. T., and Heineman, W. R., eds., *Laboratory Techniques in Electrochemistry,* Marcel Dekker, New York, 1985.

Lyons, M. E. G., ed., *Electroactive Polymer Electrochemistry,* Parts 1 and 2, Plenum Press, New York, 1996.

Pletcher, D., *A First Course in Electrode Processes,* The Electrochemical Consultancy, UK, 1991.

Rieger, P. H., *Electrochemistry,* Prentice-Hall, Englewood Cliffs, NJ, 1987.

Sawyer, D. T., Sobkowiak, A., and Roberts, J. L., *Experimental Electrochemistry for Chemists,* 2nd ed., Wiley Interscience, New York, 1994.

Smyth, M. R., and Vos, J. G., *Analytical Voltammetry. Comprehensive Analytical Chemistry,* Vol. XXVII (series editor G. Svehla), Elsevier, Amsterdam, 1992.

Wang, J., *Stripping Analysis, Principles, Instrumentation and Applications,* VCH Publications, Deerfield Beach, FL, 1984.

CHAPTER

4

BIOMATERIALS FOR BIOSENSORS

TERESA McCORMACK, GARY KEATING, ANTHONY KILLARD,
BERNADETTE M. MANNING, and RICHARD O'KENNEDY[1]

School of Biological Sciences, Dublin City University, Dublin, Ireland

4.1. GENERAL INTRODUCTION

Many key reactions in biological systems are based on recognition phenomena. For example, metabolic reactions rely on enzymes acting specifically on a molecule to convert it into a product. This product may be used directly by an organism for its growth or the product may be further acted on by enzymes with the release of energy in a usable form. The immune system has as key roles the defense against disease and the removal of nonnormal cells. Here the important recognition event is mediated by the interaction of an antigen with an antibody. The antigen (foreign material) is recognized as being "foreign," and a specific antibody is produced against it, binds to it, and mediates its removal. The blueprint for each cell is contained in its DNA (deoxyribonucleic acid). DNA contains all the information coded in a series of bases. Recognition of such sequences is of fundamental importance in the control, reading, and detection of this information. Biological systems have sensors that can detect pH, temperature, ion movements, color, odors, sizes, and other parameters. In short, biological systems have very sophisticated, sensitive, and specific sensors on which they depend, to maintain their equilibrium and respond to changes in the environment to facilitate life. The purpose of this chapter is to examine some of these systems and to see how we can exploit the biomaterials involved to develop usable biosensors for analysis.

4.2. ENZYMES

4.2.1. Introduction

Enzymes are biocatalysts that play a key role in all the biochemical reactions that make up the complexity of living systems. Enzyme-catalyzed reactions have very

[1]Author to whom correspondence should be addressed.

133

high reaction rates, typically 10^6–10^{12} times greater than uncatalyzed reactions. These reactions can normally occur under relatively mild conditions of temperature and pH. However, there are enzymes that can function under extremes of temperature and pH, and these properties are often exploited for commercial and other applications. Enzymes display a range of specificities. Some enzymes react only with a single molecule (or substrate). This is often referred to as *absolute specificity*. This may be a key factor in the selection of an enzyme for use in a specific sensor. Other enzymes may react with a variety of substrates; for example, a protease enzyme, such as trypsin, will act on proteins containing the amino acids lysine and arginine linked together.

Many enzymatic reactions require the presence of other molecules that assist the reaction. These are often referred to as *cofactors*. In many cases, the enzyme will not act in the absence of these factors. Examples of cofactors would include metals such as calcium, magnesium, or zinc. The latter metal is required for the activity of enzymes such as carboxypeptidase A. NAD (nicotinamide adenine dinucleotide) is a key organic cofactor (coenzyme) involved in many enzymatic reactions. It is often found in enzyme-based sensor systems. Often cofactors may be chemically bound, by covalent bonds, to the protein. Such molecules are called *prosthetic groups*.

Cofactors play a key role in many enzymatically catalyzed reactions. They may often be chemically changed during the course of the reaction. Such changes may be monitored as a way of following or detecting the enzymatic reaction. This may occur through changes in the optical properties of the cofactors or their oxidation–reduction changes that can be monitored, such as NAD/NADH.

4.2.2. Enzyme Structure

Enzymes are made up of amino acids bound together by peptide bonds. A typical structure of an amino acid is shown in Figure 4.1. There are over 20 amino acids with different side chains (R group). The nature of these side chains determine the properties of the amino acid. They may be hydrophobic or hydrophilic. The sequence of amino acids makes up the primary structure of the enzyme. Because of the properties of the amino acids, the structure may possess a helical structure. There are also other possible arrangements such as turns and pleated sheets. This makes up the secondary structure of a protein or enzyme and is mediated by the sequence and properties of the amino acids. The tertiary structure of a protein or enzyme refers to its three-dimensional structure. This is due to the molecule arranging itself so as to minimise non-favorable thermodynamic interactions with its environment; for instance, nonpolar amino acids tend to be sequestered away from the aqueous environment. This structure may be stabilized by hydrogen bonding, disulfide linkages, electrostatic van der Waals forces, and hydrophobic interactions. The 3D structure of the enzyme is such that part of it forms the sites

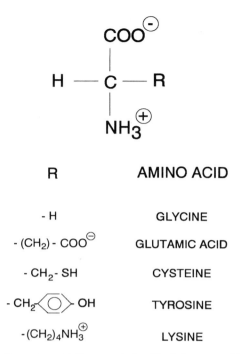

Figure 4.1. Structure of an amino acid; R represents the side-chain group, which distinguishes different amino acids.

that are directly involved in binding the substrate, the prosthetic group, and any other cofactors. In many cases, enzymes may consist of more than one subunit. Such subunits may or may not be identical. The combination and spatial arrangement of such units into a functional, active complex is known as the *quaternary structure*. The primary, secondary, tertiary, and quaternary structures are represented in Figure 4.2. It is beyond the scope of this chapter to cover this area in depth, and the reader is referred to a number of excellent standard biochemistry texts [1,2].

4.2.3.   Theory of Enzymatic Action

The catalytic activity of enzymes may be explained on the basis of transition-state theory. Suppose we examine the reaction

$$R \to P \tag{4.1}$$

PRIMARY
STRUCTURE
(amino acid sequence)

SECONDARY
STRUCTURE
(helix)

TERTIARY
STRUCTURE
(3-dimensional
arrangement)

QUATERNARY
STRUCTURE
(arrangement of
subunits)

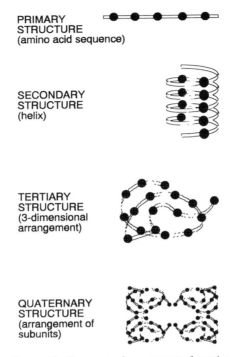

Figure 4.2. The structural arrangement of proteins.

where R is the starting material(s) or reactants and P is the product. We can rewrite this equation in terms of transition state theory as

$$R \rightarrow [\text{transition state}] \rightarrow P \qquad (4.2)$$

The rate of the reaction is proportional to the number of molecules in the transition state. In order to achieve the transition state, the molecules must have a certain energy called the *activation energy*. This can be seen more clearly by reference to Figure 4.3. Enzymes or catalysts function by lowering the free energy of activation for the reaction.

4.2.4. Enzyme Kinetics and the Michaelis–Menten Equation

Enzyme-catalyzed reactions undergo a phenomenon known as *saturation* with substrate. In Figure 4.4 it can be seen that above a certain saturating level of substrate, the velocity of the reaction reaches a maximum. This is due to the fact that the active site of the enzyme is working at full capacity—it is saturated with sub-

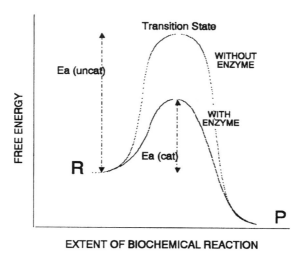

Figure 4.3. Energy diagram for enzyme-catalyzed reaction.

strate molecules. This lead Michaelis and Menten and, later, Briggs and Haldene to suggest the following theory for enzyme action and kinetics [3]:

$$E \;\;+\;\; S \;\;\underset{k_2}{\overset{k_1}{\rightleftharpoons}}\;\; ES \;\;\underset{k_4}{\overset{k_3}{\rightleftharpoons}}\;\; E \;\;+\;\; P \qquad (4.3)$$

enzyme   substrate   enzyme   enzyme   product
                                      substrate
                                      complex

The reactions are considered reversible and k_{1-4} are specific rate constants.

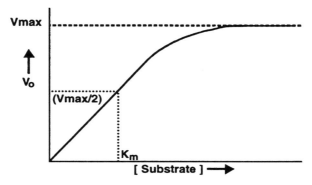

Figure 4.4. Effect of substrate concentration on the rate of an enzyme-catalyzed reaction (K_m = Michaelis–Menten constant). V_0 is the initial velocity.

If we consider the rates of formation of ES, we obtain

$$\frac{d[\text{ES}]}{dt} = k_1([\text{E}] - [\text{ES}])\,[\text{S}] \qquad (4.4)$$

where ([E]–[ES]) is the concentration of the "free" enzyme. The rate of formation of ES from E + P is small and can effectively be omitted. The rate of breakdown of ES is

$$-\frac{d[\text{ES}]}{dt} = k_2[\text{ES}] + k_3\,[\text{ES}] \qquad (4.5)$$

At the steady state, the rates of formation and breakdown of ES are equal:

$$k_1([\text{E}] - [\text{ES}])[S] = k_2[\text{ES}] + k_3[\text{ES}] \qquad (4.6)$$

or

$$\frac{([\text{E}] - [\text{ES}])[S]}{[\text{ES}]} = \frac{k_2 + k_3}{k_1} = K_m \qquad (4.7)$$

where K_m is referred to as the *Michaelis–Menten constant* and

$$\frac{[\text{E}][\text{S}] - [\text{ES}][\text{S}]}{[\text{ES}]} = K_m \Rightarrow \frac{[\text{E}][\text{S}]}{[\text{ES}]} - [\text{S}] \qquad (4.8)$$

from which

$$[\text{ES}] = \frac{[\text{E}][\text{S}]}{K_m + [\text{S}]} \qquad (4.9)$$

The initial rate (v) of an enzymatic reaction is proportional to [ES], therefore:

$$v = k_3[\text{ES}] \qquad (4.10)$$

When the concentration of substrate is so high that the enzyme is saturated, the rate of the reaction depends on [E]:

$$V_{\text{max}} = k_3[\text{E}] \qquad (4.11)$$

therefore

$$v = k_3[ES] = k_3 \frac{[E][S]}{K_m + [S]} \qquad (4.12)$$

therefore

$$\frac{v}{V_{max}} = \frac{k_3[E][S]/(K_m + [S])}{k_3[E]} \qquad (4.13)$$

therefore

$$v = \frac{V_{max}[S]}{K_m + [S]} \qquad (4.14)$$

This is usually referred to as the *Michaelis–Menten equation*.

A special case exists when $v = \frac{1}{2}V_{max}$:

$$\frac{V_{max}}{2} = \frac{V_{max}[S]}{K_m + [S]} \qquad (4.15)$$

$$K_m + [S] = 2[S] \qquad (4.16)$$

or

$$K_m = [S] \qquad (4.17)$$

Here, K_m is the substrate concentration at which the velocity is half-maximal. This is illustrated in Figure 4.4.

Michaelis–Menten kinetics apply to enzymes in solutions. However, for the most part, enzymes used in sensor systems are immobilized, adsorbed, or entrapped onto a surface (e.g., an electrode). This has a number of implications:

1. The accessibility of the active site may be somewhat limited.
2. The immobilization/adsorption–entrapment step may impose diffusion limitations on the substrate and/or the product of the reaction and may also modify the kinetic characteristics of the enzyme.
3. The stability of the enzyme may be altered (ideally it should be increased).

The implications of these conditions and kinetic equations that partially describe them have been extensively reviewed [4]. Factors to take into consideration include the presence or absence of a membrane or medium that can effect diffusion, pH effects, and the thickness of the enzyme layer immobilized. However, some of these effects will be of less significance where the enzyme is used as a label (e.g., on an antibody or antigen) or where it is in free solution.

There are also some sensor systems which may involve the use of enzymes that react specifically with substrates that inhibit their activity, for example toxins that inhibit enzymes such as acetylcholinesterase involved in neurochemical reactions. Such competitive inhibition can be described as follows:

$$E \quad + \quad I \quad \underset{k_2}{\overset{k_1}{\rightleftharpoons}} \quad EI \tag{4.18}$$

enzyme   inhibitor   enzyme inhibitor
complex

Hence, the amount of free (i.e., uncomplexed) enzyme now equals $E - [ES] - EI$. However, EI cannot break down to form E. Therefore, an inhibition constant K_I, can be defined as

$$K_I = \frac{k_2}{k_1} \tag{4.19}$$

Since [from (4.14)]

$$v = \frac{V_{max}[S]}{K_m + [S]}$$

and

$$K_I = \frac{k_2}{k_1} = \frac{[E][I]}{[EI]} \tag{4.20}$$

then

$$v = \frac{V_{max}[S]}{K_m(1 + [I]/K_I) + [S]} \tag{4.21}$$

or

$$\frac{1}{v} = \frac{K_m(1 + [I]/K_I)}{V_{max}[S]} + \frac{1}{V_{max}} \tag{4.22}$$

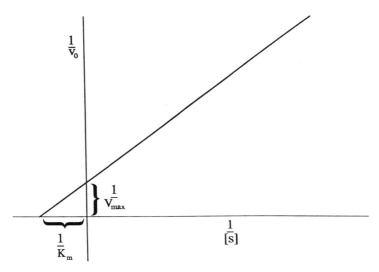

Figure 4.5. Lineweaver–Burke plot, which allows an accurate determination of V_{max}. The plot is $1/V_0$ = $(K_m/V_{max})(1/[S]) + (1/V_{max})$; therefore slope = K_m/V_{max} (from $y = mx + c$, where m = slope). (After Lehninger [3].)

This reciprocal transformation, known as a *Lineweaver–Burke equation,* is of the form $y = mx + c$. A plot of the Lineweaver–Burke equation is shown in Figure 4.5.

Many biosensor systems exploit enzymes, and some of the properties that make them useful are outlined in Table 4.1. Table 4.2 lists a range of different ways in which enzymes are used in sensor systems. Indeed, it shows both the diversity of the application and the techniques used to monitor the enzymatic reactions. These include optical, fluorescence, luminescent, electrochemical, calorimetric, and pH-based systems.

Since many enzymes have high specificity for their substrates, this property has been widely exploited in sensors for the determination of specific analytes. Examples of these are given in Table 4.3. While this is by no means a comprehensive list, it clearly indicates that the use of enzymes is well established in analysis using sensors (see Hall [5]).

4.3. ANTIBODIES

4.3.1. Introduction

The primary function of an animal's immune system is to defend it against any foreign molecules that may invade its body. Viruses, bacteria, fungi, and parasites can

Table 4.1. Enzyme Properties of Advantage in Sensors

1. High specificity for substrate or analyte
2. Reusable
3. Well-characterized mechanism of action
4. Many ways of monitoring reactions catalyzed
5. Well-developed chemistry and methodologies for immobilization
6. Synthetic substrates available that provide products, following enzymatic conversion, that can be easily detected; facilitates direct determination of enzymatic activity
7. Many analytical formats available (e.g., direct, indirect, competitive, noncompetitive)
8. Stable forms of enzymes available; may be produced from thermophilic organisms or generated by chemical or genetic mutation; a very active area of development
9. Commercial systems based on enzymes well established (e.g., for determination of glucose, cholesterol, and other biomolecules)
10. Relatively cheap to produce and purify; now feasible to produce less readily available and expensive enzymes using genetic engineering methods; also allows the design of enzymes with special characteristics (e.g., greater stability; altered specificity)
11. Many cell- and tissue-based sensors are enzyme-based

kill their host if they are allowed to multiply unchecked. In protecting the host, the immune system essentially surveys the body and eliminates any foreign molecules it finds therein by a series of complex mechanisms. The immune system has two main divisions: the innate immune system and the adaptive immune system. The innate immune system constitutes the first line of defense and is mediated by cells in the body that respond nonspecifically to foreign molecules. These cells include *phagocytes,* which engulf foreign particles and destroy them; and *natural-killer*

Table 4.2. Some Examples of the Utilization of Enzymes in Sensor Systems

Determination of specific analytes (e.g., amino acids, sugars, cholesterol, glucose)

As a label for analytes in competitive assay systems or as a label for antibodies (e.g., horseradish peroxidase, alkaline phosphatase, urease, glucose oxidase)

In optical systems (e.g., formation of NAD/NADH with measurement of associated changes in absorbance at 340 nm)

In oxygen-linked assay systems (e.g., glucose determination using glucose oxidase)

In redox systems (e.g., with flavine or pyridine-linked dehydrogenases or cytochromes)

For amplification of weak signals through a series of linked enzyme-based reactions

In the generation of fluorescence and/or luminescence

Through the production of pH changes due to the catalytic effect of the enzyme on a substrate

Generation of heat due to the reaction catalyzed—may be detected using thermistors

In catalytic antibody (abzyme) systems

Table 4.3. Specific Enzyme-based Sensors

Component	Enzyme
Glucose	Glucose oxidase
Cholesterol	Cholesterol oxidase
Creatine	Creatinase
Penicillin	Penicillinase
Urea	Urease
Dopamine	Monoamineoxidase
Acetylcholine	Acetylcholinesterase
Ethanol	Alcohol dehydrogenase
Lactic acid	Lactic acid dehydrogenase
Nitrite	Nitrate reductase

cells, which bind to the foreign molecules and kill them. The adaptive immune system, however, produces a specific reaction to invading molecules, involving the production of molecules that specifically recognize the invading molecule on the basis of its structure.

B-Lymphocytes, the cells that mediate the adaptive immune system, produce specific binding proteins, called *antibodies,* which bind to the foreign molecules (termed *antigens*), causing them to be eliminated. Each B-lymphocyte in the body is inherently capable of making an antibody, which is located on its surface as an antigen receptor. The invading antigen binds only to B-lymphocytes having antibodies on their surfaces that react specifically with it, and in so doing stimulate the B-lymphocytes to divide and mature into antibody producing cells and memory cells. The "clonally" produced antibodies bind to the antigen and in doing so, activate other parts of the immune system to remove the antibody-antigen complex from the body. Memory cells, which are differentiated lymphocytes, enter the lymphatic and blood circulation systems of the body. From here, they patrol the tissues of the body. This means that if a reinfection should occur through a point of entry far distant from the previous infection, migratory immune memory lymphocytes could soon deal with it. In the case of invading pathogenic molecules, these memory cells can prevent the development of the clinical symptoms of disease by eliminating the pathogen quickly.

The specific molecular recognition and interaction between antibodies and antigens can be exploited in vitro and forms the basis of a large number of clinical diagnostic tests. Antibodies raised against specific molecules can be employed to detect these molecules by labeling them with enzymes, fluorophores, or radioisotopes. Such immunoassay systems will be discussed in detail. Immunosensors, based on electrochemical and optical detection systems, also utilize antibody–antigen interactions as the molecular sensing element. Case studies on the use of antibodies in biosensors are outlined.

4.3.2. Antibody Structure

The basic structure of an immunoglobulin or antibody is shown in Figure 4.6. Immunoglobulin molecules are glycoproteins of molecular weight between 150,000 and 160,000. They consist of four polypeptide chains, two identical heavy (H) chains, and two identical light (L) chains. These heavy and light chains are held together via disulfide bridges and noncovalent forces, the arrangement of which are shown.

The heavy and light chains can be divided into constant (C) and variable (V) regions. Each heavy chain (m.w. 55,000) has one V region or domain (V_H) and

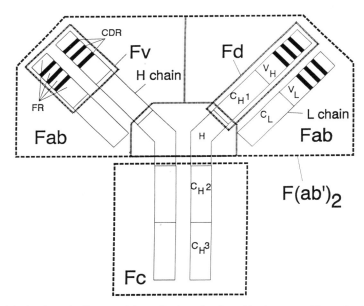

Figure 4.6. A schematic diagram of the antibody molecule and its fragments. The molecule can be subdivided into two parts. The Fc (crystalline fragment) contains the antibody effector functions. The Fab (antibody-binding fragment) consists of the heavy (H) and light (L) variable (V) chains (V_H and V_L) and the constant (C_H1 and C_L) chains. When joined to its neighboring Fab via the hinge (H) region, this is designated $F(ab')_2$. Other segments are the Fv (variable fragment) consisting of the V_H and V_L chains and the Fd fragment, which contains the V_H and C_H1 chains. Each V region contains three complementarity-determining regions (CDRs), bordered by four framework (FR) regions. The Fc and $F(ab')_2$ portions are bounded by the dashed line and the vertical line in the center. The Fab fragments and the hinge region are separated by the bold line. Fv and Fd fragments are also bounded by the bold line.

normally three C regions (C_H1, C_H2, C_H3), while the light chain (m.w. 25,000) has one V region (V_L) and one C region (C_L). The variable regions of one heavy and one light chain associate together to form an antigen binding site. The rest of the IgG class of immunoglobulin molecule is composed of three constant regions or domains. The constant region of the light chain, C_L, can be either of type kappa or lambda and is associated with the constant region of the heavy chain, termed C_H1. Within any immunoglobulin molecule both light chains are of one type as hybrids do not occur naturally. The C_H2 and C_H3 are linked via disulfide bridges with the second identical heavy chain to form the Fc (crystalline) portion of the immunoglobulin molecule.

There are five classes of immunoglobulin found in serum, IgG, IgM, IgA, IgD, and IgE. The classes are distinguished from one another by the type of heavy chain found in the molecule and the number of structural units it possesses. The IgG class is involved in the secondary immune response. The IgG antibody contains only one structural unit and is bivalent, having two "arms" containing sites that can bind to a particular part of the antigen molecule called the *epitope*. The hinge region is found between the C_H1 and C_H2 domains and allows both lateral and rotational movement of the antigen binding sites, allowing the antibody more flexibility to interact with different antigen conformations.

4.3.3.   Functions of Antibody Domains

The individual domains of an IgG molecule have different physiological functions. These functions were elucidated using enzymes that "digest" the antibody into various combinations of its individual domains. One such enzyme, papain, cleaves the IgG molecule at the hinge region, dividing the immunoglobulin into three fragments, two identical Fab fragments and the Fc receptor fragment. It was discovered that the Fab fragment was involved in antigen binding. Other enzymes, such as pepsin, have been used to cleave the immunoglobulin into fragments. Digestion by pepsin produces two major fragments: the $F(ab')_2$ fragment, which is roughly equivalent to the two Fab fragments joined at the hinge region; and the pFc' fragment, corresponding to the Cγ3 domain of the molecule. A degraded fragment of the C_H3 region—the Fc' fragment—is also produced by digestion with pepsin. Acid-treated Fc fragments, which have been acted on by the enzyme trypsin, are degraded to the C_H2 domain. It is also possible to produce an Fv fragment, consisting of the variable light- and heavy-chain domains. The functions of the various fragments were elucidated, and it was discovered that the Fc portion of the molecule has important physiological roles in binding to effector units of the immune system such as complement, phagocytes, and neutrophils. These effector molecules initiate the removal of the antigen, to which the antibody is bound, by activation of various complex mechanisms.

The Fc portion of the molecule is not required for antigen binding, and therefore the $F(ab')_2$ can be utilized for applications where the physiological effects of the immune system are not required. In immunoassay systems, one advantage of using such modified antibodies is a reduction in nonspecific binding. The term Fv represents the portion of the IgG molecule made up by the V_H and V_L chains. It was discovered that this portion of the molecule could bind to the antigen in the absence of the rest of the molecule. Further studies indicated that only short segments of the amino acid sequence in the peptide chain found in the variable region were actually involved in antigen binding. These were called the *complementarity-determining regions* (CDRs). The rest of the variable region was made up of highly conserved framework regions (FRs). The specificity for antigen binding is determined by the three-dimensional structure of the antibody combining site, and the precise interaction depends on the amino acid residues that react with the antigen.

4.3.4. Production of Antibodies

Antibodies can be produced against any molecule capable of eliciting an immune response in a host animal. For an immune response to be produced against a particular molecule, a certain molecular size and complexity is necessary; proteins with molecular weights greater than 5000 daltons are almost invariably immunogenic, and so, proteins taken from one species will elicit an immune response when injected into another species. Other *immunogenic* moieties include tissue extracts, cells, viruses, bacteria, and spores. Smaller molecules such as drugs, lipids, steroid hormones, peptides, and antibiotics will not elicit an immune response by themselves but must be bound to a larger immunogenic molecule called a *carrier molecule*. Any substance that binds specifically with an antibody is termed an *antigen*. An *immunogenic* molecule, however, possesses several distinct locations or *antigenic determinants* on its surface, each of which is capable of eliciting the production of antibodies. Therefore, a single immunogenic molecule will have antibodies directed against several sites on its surface. Nonimmunogenic substances can be made immunogenic by linking them to larger immunogenic molecules (e.g., proteins). In producing antibodies to the larger molecule, the immune system will also produce antibodies to the molecules conjugated to its surface, as these molecules now appear as antigenic determinants on the surface of the larger molecules. Any substance that is not immunogenic but is antigenic in that antibodies can be raised against it in the manner described above is termed a *hapten*.

Rabbits, goats, sheep, guinea pigs, and fowl are examples of the animal host species often used for polyclonal antibody production; rats and mice are used mainly for monoclonal antibody production. The choice of animal is based on several requirements, including the amount of antiserum required and the source of antigen. For polyclonal antiserum production, rabbits are a good choice. The best immune response is obtained when the animal chosen does not express the antigen or any closely related protein. All immune systems display self-tolerance in that they do not recognise their own molecules as foreign, thus protecting themselves from autoimmune damage. Therefore, when raising antibodies to highly conserved mammalian proteins, fowl are often a good choice of animal and will produce strong immune responses. Where possible, several animals should be used for immunisations, since even with genetically identical animals, a single preparation of antigen will stimulate different B-lymphocytes producing different antibodies against it.

Prior to immunization, the antigen, especially when in soluble form, should be mixed with nonspecific stimulators of the immune system called *adjuvants*. Soluble antigens are quickly removed from the body by the innate immune system. Adjuvants act by prolonging the exposure of the antigen to the immune system, thereby ensuring a secondary immune response and IgG production. Adjuvants also act by increasing the efficiency of antigen presentation, enhancing immunogenicity of the antigen and increasing the average affinity and avidity of the antibody response. The most common adjuvant used for soluble antigens is Freund's complete adjuvant and Freund's incomplete adjuvant. *Freund's complete adjuvant* is composed of a mixture of a mineral oil, an emulsifier and a suspension of heat-killed *Mycobacterium tuberculosis*. Immunogens are emulsified in the adjuvant to form a stable oil-in-water suspension. The mycobacterium stimulates the cells of the immune system, while the oil-in-water suspension serves to create a depot of immunogen that prolongs the release of the immunogen so that it only slowly becomes available to the immune system. This extension of the immune response also ensures the production of the IgG class of immunoglobulin. Freund's complete adjuvant is only used for the initial immunization, and *Freund's incomplete adjuvant* (no mycobacterium) is used for subsequent booster injections, due to its reduced pathogenic effects.

Immunogens can be injected into several sites in the animal (e.g. subcutaneously, intramuscularly, intraperitoneally, intravenously, intradermally, or intrasplenically). The choice of route depends on the volume that must be delivered, the physical nature of the injection, the species of animal, how quickly the immunogen should be released into the lymphatics or circulation, and the stage of the immunization routine.

4.3.4.1. *Polyclonal Antibodies*

When an animal is immunized with an immunogen, a mixture of antibody molecules is produced. They will all bind the antigen, but do not all bind at the same sites or with equal strengths. This mixture of polyclonal antibodies is composed of antibodies produced by many different B-lymphocytes. The general immunization schedule for the production of polyclonal antibodies involves immunizing at 3–4 week intervals. Following the second immunization, the animal may be bled

Table 4.4. Commonly Used Methods for Purification of Immunoglobulin from Serum (Polyclonal Antibodies) and Ascitic Fluid or Culture Supernatent (Monoclonal Antibodies)

Purification Method	Basis of Separation	Level of Purification	Reference
Salt fractionation	Increasing amounts of salt induces protein–protein interactions, resulting in precipitation	Crude—all immunoglobulin protein precipitated	6
Ion-exchange chromatography	Charged immunoglobulins bind to oppositely charged beads; eluted by changing environmental conditions so as not to favor the interaction	Good—method can be manipulated to selectively purify different classes and subclasses	7
Protein A chromatography	Immobilized protein A interacts with IgG Fc receptor; eluted by changing environmental conditions so as not to favor the interaction	Good—method selective for IgG class of immunoglobulin	8
Affinity chromatography	Immobilized antigen interacts with its corresponding antibody; eluted by changing the environmental conditions so as not to favor Ab–Ag binding	Excellent—only specific antibody isolated	9

between 10 and 14 days after the injection. Repeated immunizations increase the titer of the antibody circulating in the serum. High levels of antibody remain in the serum for 2–4 weeks following immunization, and sufficient time should be allowed between injections to allow the circulating level of antibody to decrease; otherwise the newly injected antigen would be cleared rapidly.

Antisera for specific antigen studies require that other unwanted nonspecific immunoglobulins be removed. Depending on the degree of purification and intended use of the purified antibody, different methods of purification are employed (Table 4.4). The most commonly used initial purification step for immunoglobulins is salt fractionation. Specific antibody can be isolated by affinity chromatography, where the antigen to which the antibody has been raised is immobilized onto

Sepharose beads. The antibody binds specifically to the antigen, and following removal of the unbound protein, the antibody is recovered by breaking the antibody–antigen bond. This is generally achieved by lowering the pH of the buffer. Protein A, isolated from the cell walls of the staphylococcal strain of bacteria, binds to the Fc portion of the IgG class of immunoglobulin molecule and can be used for its purification in a similar manner to affinity chromatography. Immunoglobulin, not specific antibody is isolated in this way. Other methods of purification include ion-exchange chromatography, which separates proteins on the basis of their net charge at a certain pH.

4.3.4.2. *Monoclonal Antibody Production*

Polyclonal antisera are heterogenous and may be unsatisfactory for certain analytical applications. A preparation of identical antibodies that all bind to the same epitope on the antigen surface with the same strength of binding offers a potentially superior immunoanalytical tool. Kohler and Milstein [10] reported the method for production of such monoclonal antibodies, the scheme for which is illustrated in Figure 4.7.

As described earlier, every B-lymphocyte is predetermined to make one particular antibody molecule, which it expresses on its surface as a receptor for antigen. Only on binding of this receptor by antigen is the cell stimulated to differentiate and produce a clone of specific antibody-secreting cells. If these cells could be continuously propagated in culture, they would provide a constant supply of homogenous monoepitope-specific antibodies or monoclonal antibodies. Myeloma cells, or malignant B-lymphocytes, are immortalized cells and can be grown in vitro continuously, whereas normal B-lymphocytes have a finite lifespan in culture. If a myeloma cell and a normal B-lymphocyte are fused together to form a hybridoma, we now have a source of continuously produced monoclonal antibodies. The myeloma cells chosen for cell fusions are derived from common ancestor lines that have lost the ability to produce any immunoglobulin of their own.

Mice or rats are immunized and screened for antibody production as for polyclonal antibodies. The spleen of the animal is removed, as the spleen is a rich source of B-lymphocytes. The cells of the spleen are fused with the myeloma cells using PEG (polyethylene glycol), a fusogen or promoter of cell fusion. Following fusion, however, spleen, myeloma, and hybrid cells are present. Spleen cells die off naturally, but both the myeloma cells and the hybridomas can grow continuously in culture. The myeloma cells must therefore express a feature that makes it possible to grow the hybrids selectively in the presence of the parent myeloma. To achieve this, the myeloma cells are mutated so that they lack the enzyme, hypoxanthine guanine phosphoribosyl transferase (HGPRT). This enzyme is required to incorporate hypoxanthine into the cell.

In order to grow, cells must synthesize RNA and DNA. The principal biosyn-

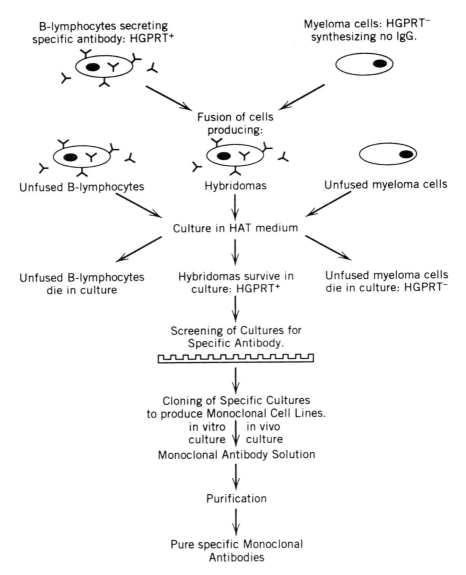

Figure 4.7. Schematic diagram of the production of murine monoclonal antibodies. B-Lymphocytes secreting antibody are fused with myeloma cells that lack the enzyme, hypoxanthine guanine phosphoribosyl transferase (HGPRT), and the fusion products are cultured in medium containing HAT (hypoxanthine, aminopterin, thymidine). The surviving hybridomas are screened for specific antibody production. Specific cultures are cloned until a single antibody clone is obtained.

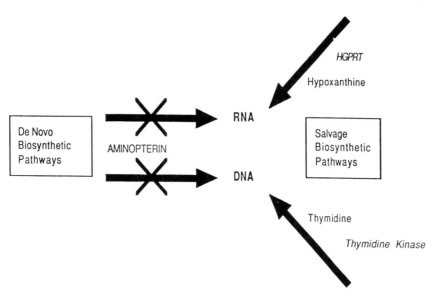

Figure 4.8. Principle of HAT selection system. The principal nucleic acid biosynthetic pathways are blocked by the drug aminopterin. However, if the cells are provided with hypoxanthine and thymidine, they can synthesize DNA and RNA via "salvage" biosynthetic pathways employing the enzymes, hypoxanthine guanine phosphoribosyl transferase (HGPRT) and thymidine kinase. Cells which have been mutated so that they lack either of these enzymes cannot synthesize nucleic acids and die.

thetic pathways for RNA and DNA are blocked by the drug aminopterin. However, if provided with hypoxanthine and thymidine, cells can synthesize these nucleotides via "salvage" biosynthetic pathways using the enzymes HGPRT and thymidine kinase. This process is outlined in Figure 4.8. Hybrid cells can be selectively grown in the presence of the parent myelomas by culturing them in medium containing HAT (hypoxanthine, aminopterin, and thymidine). Myeloma cells, lacking HGPRT, die off because they cannot use the principal or salvage pathways to synthesize DNA. Hybrids, however, may have the enzyme HGPRT from the spleen parent and, therefore, can multiply in HAT medium, using the salvage pathways. Following a successful fusion, the cultures are screened for specific antibody production and the selected cultures are cloned. Screening methods should be well worked out prior to a fusion and should be fast, sensitive, reliable, and relevant to the format in which the antibody is intended for use. Methods for screening include enzyme-linked immunosorbent assays (Section 4.4.4). The process of cloning allows the selection of a positive hybrid which is derived from a single cell; thus, any colonies that are composed of a mixture of cell types can be separated into pure populations. Cloning is achieved by seeding single cells into culture wells and allowing them to proliferate into colonies. Other methods of cloning include using semisolid agar or fluorescence-activated cell sorting. Specific antibody-producing hybridomas can be grown on a large scale as ascitic tumors in animals or grown continuously in culture. Finally, the monoclonal antibodies may be purified, and the method of choice will depend on the intended use of the antibodies, the level of purity required, the class of antibody involved, and the source of antibody. Purification methods are similar to those used for polyclonal antibodies.

Human monoclonal antibodies are required for many in vivo immunotherapeutic applications [11]. Such antibodies can be produced in vitro [12]. The method of production is similar to that used for the production of murine monoclonal antibodies, except that for ethical reasons, human lymphocytes must be removed and immunised in vitro. Another problem that hampers the production of human monoclonal antibodies is the lack of available human myeloma partner cells. An alternative strategy to the somatic hybridisation approach is to transform antibody-producing cells using the Epstein–Barr virus (EBV). No fusion partner is required as the human B-lymphocytes are immortalised and can grow continuously in culture, although fusion is often used in conjunction with EBV transformation. An alternative to this is to "humanize" murine monoclonal antibodies or produce chimeric antibodies that have murine-derived antigen binding sites attached to the constant structural elements (Fc portion) of the human antibody. There are a number of drawbacks to these systems that prompted new developments in antibody engineering.

4.3.4.3. Combinatorial Phage Display Libraries

Although the production of monoclonal antibodies of animal origin is now commonplace, a suitable equivalent for human monoclonal antibody production has not proved possible. There are few myeloma fusion partners of human origin, and all these secrete some antibody fragments (Bence–Jones proteins). This results in a heterogeneous antibody population following fusion. Human hybridomas are also inherently unstable and may cease to produce some antibody chains [13]. Attempts to immortalize human B lymphocytes using the Epstein–Barr virus have had only limited success as these hybridomas are also very unstable and cease antibody production after a relatively short period [14]. Also, immunization of humans raises complex ethical considerations. To circumvent these problems, a system known as *combinatorial phage display* has been developed. This allows the in vitro production of antibody Fab fragments of human origin with defined specificity and affinity.

A popular combinatorial phage display system is that of Barbas et al. [15]. This

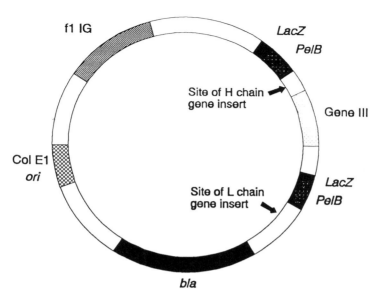

f1 IG

LacZ
PelB

Site of H chain
gene insert

Gene III

Col E1
ori

Site of L chain
gene insert

LacZ
PelB

bla

Figure 4.9. The phagemid vector pComb3. The ColE1 origin of replication (*ori*) and the antibiotic resistance gene (*bla*) allow for the stable replication and maintenance of the phagemid in *Escherichia
coli*. The phage f1 intergenic (IG) region allows the production of phage proteins and allows the
phagemid to be packaged in phage particles in the presence of helper phage. Sites are present for the
insertion of the light and heavy chain antibody genes. The site for insertion of the heavy chain gene is
joined to the phage gene *gIII*. The *pelB* sequences ensure the export of the produced heavy- and light-
chain proteins to the periplasm of *E. coli*, and the level of production of these proteins can be increased
by induction of the *lacZ* promoter.

is based on the phagemid vector pComb3 (Figure 4.9). A phagemid is a DNA molecule that possesses the properties of a plasmid and a phage (hence *phagemid*). A
plasmid is a small circular DNA molecule that is capable of replicating in a bacterial cell. The phagemid *pComb3* is able to replicate in the bacteria *Escherichia
coli* (*E. coli*) as the phagemid possesses a replication origin (*ori*) that is recognized
by the DNA replication machinery in this organism. Cells containing the phagemid
can be selected from those that do not by the presence of an antibiotic resistance
gene *bla*. A phage is a bacterial virus and it is capable of infecting and replicating
in bacterial cells. *Escherichia coli* can be infected by several phages, a family of
which is the single-stranded phages M13, f1, and fd. The phagemid has a region
of DNA taken from phage f1 called the intergenic (IG) region. This allows the
phagemid to act as viral DNA in certain circumstances. The phagemid also contains regions where antibody chain genes can be inserted. The region for the insertion of the heavy-chain gene is adjacent to a phage gene *gIII*, also taken from

M13. This gene codes for a phage coat protein pIII. (For an outline of plasmid and phage vectors, see Old and Primrose [16].)

The process of producing a Fab antibody library using the pComb3 vector is shown in Figure 4.10. The genetic material from phage display comes from human B-lymphocytes. These can be derived from bone marrow or peripheral blood. It is not essential for the lymphocyte donor to have been immunized. Sufficient lymphocytes (10^8 cells) are used to make it probable that every antibody specificity is represented in the sample.

B-Lymphocytes producing antibody molecules contain messenger RNA (mRNA) coding for these proteins. This can be converted to complementary DNA (cDNA) and amplified to sufficient quantities by a technique called the polymerase chain reaction (PCR) [17]. The heavy-chain genes are inserted into the vector first by cutting and rejoining (ligating) the DNA. The phagemid is grown to large numbers in *E. coli* and, in a similar process, the light-chain genes are inserted. The phagemid is now placed back in *E. coli* along with a helper phage. The helper phage is derived from M13 and produces all the proteins necessary to make a phage particle, but its DNA does not become part of the phage. The final result is a phage particle that contains the phagemid with DNA at its core. Phage proteins are produced from helper phage genes and heavy and light antibody chains are produced from pComb3. The heavy chain is connected to the phage coat protein pIII and is expressed along with pIII on the surface of the phage. The light chain associates naturally with the heavy chain during phage assembly. These phage particles, now carrying an antibody fragment on their surface and the DNA encoding these antibody fragments within the phage particle, are secreted into the liquid growth medium. Phage carrying Fab specific for the antigen of interest can be selected from the rest of the population using affinity selection (panning). In this, the antigen is linked to a solid support. The mixed phage population is added, and phage carrying Fab specific for the antigen binds selectively to it. Nonspecific phage particles can be washed away, leaving specific phage behind. This is normally carried out 3 times with subsequent regrowth of the phage in *E. coli* each time.

Figure 4.10. Producing a combinatorial phage display library using the pComb3 phagemid vector. Messenger RNA (mRNA) from human B-lymphocytes is converted to DNA, and heavy (H) and light (L) antibody chain gene sequences are amplified by the polymerase chain reaction (PCR). The H-chain genes are inserted into pComb3 and grown in *Escherichia coli* (*E. coli*). The phagemid is repurified, and L-chain genes are then inserted. The phagemid is again inserted into *E. coli* and grown along with helper phage. This results in H- and L-chain proteins expressed on the surface of the phage particles by the fusion with the H chain with the phage coat protein pIII. The phage with the required specificity and binding strength (affinity) can be enriched by affinity selection (panning). Here, the antigen of interest is attached to a solid support such as a microtiter plate. High-affinity phage will bind, and those that do not can be washed away. The remaining phage can be grown again in *E. coli*. This process can be repeated to improve the enrichment. When Fabs of suitable affinity have been obtained, soluble Fab molecules can be produced by removing the gene *gIII* from the phagemid. Fab is now secreted into the surrounding medium and can be easily purified.

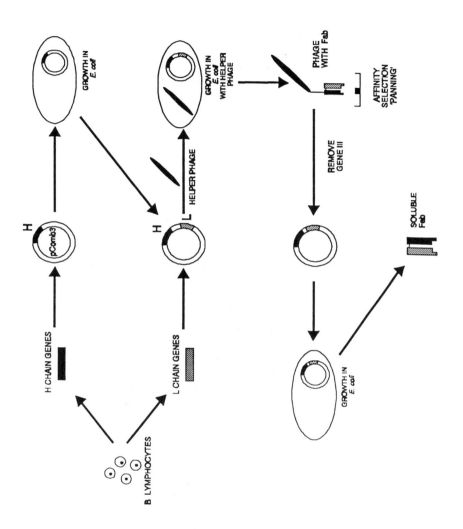

Soluble Fab antibody molecules can be produced by removing the gene *gIII* from the phagemid and growing the phagemid in *E. coli* without helper phage. The phagemid now propagates as a plasmid and produces light- and heavy-chain proteins. The level of production of these proteins can be increased by induction of the *lacZ* promoter, which increases the rate of transcription of these genes. The *pelB* sequences ensure that the light- and heavy-chain proteins are transported to the *E. coli* periplasm, where they recombine to form intact Fab fragments.

The phage display system has the capability to clone the entire human immunological repertoire, meaning that it can generate all possible combinations of antibody produced by the human body. This provides the possibility of producing therapeutically useful antibody molecules and a range of novel antibodies of defined specificity that could be utilized in sensor systems.

4.3.5. Detection of Antibody–Antigen Binding

As discussed earlier, antibodies are highly specific binding agents that can be produced to bind a very wide range of compounds whether naturally occurring or not. This specific binding event can be used to separate, identify, and quantitate the substances to which they bind. Antibodies have found many applications as qualitative and quantitative analytical tools. They have been used in diagnostic, therapeutic, preparative, and quantitative applications where the analyte is the antigen (or antibody) itself. The analysis of biochemical compounds—both specific identification and the sensitive measurement of their concentrations in given biological materials—is commonly achieved through *immunoassays.*

The detection of antibody–antigen binding was quite limited until the mid-1960s. Their interaction could only be detected by time-consuming, insensitive systems that were difficult to quantitate. Most assays were based on the formation of precipitates resulting from antibody–antigen complexation, which were detected by nephelometry, centrifugation, or immunoelectrophoresis (Table 4.5). The development of methods for chemically coupling enzymes to antibody molecules so that both the enzyme and the antibody retain their biological activity and the realization that immunoglobulins could be immobilised to solid surfaces in functional form led to the concept of the enzyme-linked immunosorbent assay (ELISA).

Initially, enzyme and fluorescently labeled antibodies were used qualitatively for the localization of antigens in histological samples. In ELISAs, however, the quantitation of antigens is facilitated by the enzyme labeling of antibodies and antigens as they allow detection of immunocomplexes formed on a solid phase, because the fixed enzyme, once washed free of excess reagents, and on subsequent reaction with its substrate, can yield a colored product that can be measured optically. The optical density of this colored product can be related to antigen concentration.

Immunoassays can be categorized as homogenous or heterogenous. In a het-

Table 4.5. Some Applications of Polyclonal and Monoclonal Antibodies

Antibody Method	Example	Principle	Polyclonal–Monoclonal	Reference
Precipitation	Immunoelectrophoresis	Ag separated by electrophoresis; diffusion into Ab gel to form precipitin arc	Polyclonal only	18
Agglutination	Passive hemagglutination	Immobilized Ab agglutinates cells having surface antigen in microtiter plate	Polyclonal–monoclonal	19
Enzyme label methods	ELISA (two-site)	Solid-phase immobilized antibody captures antigen; detected using Ab–enzyme conjugate	Polyclonal–monoclonal	20
Radiolabel methods	Immunoradiometric assay	As for ELISA	Polyclonal–monoclonal	21
	Western blotting	Antigen separated by electrophoresis—transferred to nitrocellulose and stained using labeled Ab	Polyclonal–monoclonal	22
Fluorescent label methods	Immunohistochemistry	Fluorescently labeled Ab used for cell and tissue staining	Polyclonal–monoclonal	23
	Fluorescence-activated cell sorting	Fluorescently labeled Ab used to bind cell followed by cell population separation	Monoclonals mainly; polyclonals as second Ab	24

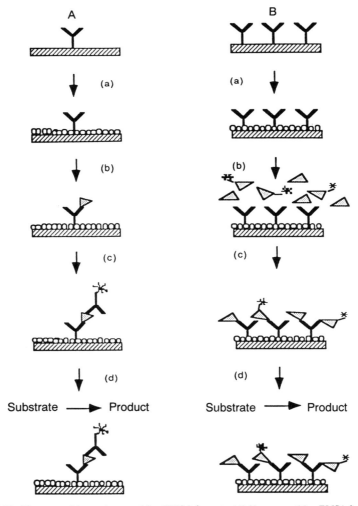

Figure 4.11. Noncompetitive and competitive ELISA formats. (*A*) *Noncompetitive ELISA format:* (*a*) Antibodies attached to solid phase. Nonspecific binding sites blocked with nonspecific protein (e.g., bovine serum albumen or gelatin). (*b*) Antigen-containing solutions incubated in the wells. Antigen binds to antibodies specifically and unbound material removed by washing. (*c*) An excess of enzyme(*)-labeled antibody incubated in the wells. This second antibody binds to a different isotope on the antigen surface than the immobilized antibody. Unbound labeled antibody removed by washing, leaving a three-layer immunocomplex attached to the solid support. This assay is often referred to as a "sandwich" immunoassay. (*d*) Substrate for the enzyme added and the optical density of the product measured after an appropriate period of incubation. The enzymatic activity is proportional to the concentration of the antigen bound to the solid surface. *Alternatively,* if the second antibody is expensive or in short supply, an *anti*-species IgG antibody can be labeled with the enzyme. Following formation of the three-layer immunocomplex, the enzyme-labeled *anti*-species IgG binds to the second antibody and is detected in the usual manner. (*B*) *Competitive ELISA format:* (*a*) Antibody attached to the solid

erogenous assay, a separation step is required where the free antibody or antigen is separated from the solid-phase-bound complexed antibody–antigen. The quantifiable signal from the immunocomplex does not differ from the free entity and, therefore, without this separation step, interference would occur. In a homogenous assay, this separation is not required. The binding of the antibody and antigen results in a change in the activity of the label. The signal obtained from the label involved in the binding reaction is altered and can be distinguished from the label that is not involved in the immunoreaction. This change in activity can be detected without a separation step and without the problems of interference.

Immunoassays can be further divided into competitive and noncompetitive (Figure 4.11). The former is based on a competitive equilibrium between an excess of labeled and unlabeled antigen (or antibody), the analyte to be determined, for a limited amount of its binding partner. An indirect relationship results since the more analyte present, the less of the labeled species that will bind. In a noncompetitive assay, the analyte to be measured complexes with its binding partner, which is immobilized on a solid phase. A second labeled antibody with specificity for a different epitope on the analyte binds to this complex and is measured using the appropriate substrate for the enzyme. A direct relationship results here.

For many applications, the information required from an immunoassay is a "yes/no" answer, specifically, whether the antigen is present in the sample or above a certain threshold value. References, both negative and positive controls, must be included in the assay to validate the results. For quantitation of antigen or antibody, a reference or standard curve must be prepared, because the relationship between the amount of color or fluorescence produced with antigen concentration may not be linear. In fact, data from ELISAs is usually plotted on linear–logarithmic paper. So long as reference material is handled exactly the same as the "unknown" test sample, the reference curve can be used to determine the concentration of the antigen. For accuracy the test sample should be diluted so that it falls on the most linear part of the reference curve.

Standardization of assays performed with polyclonal antibodies can be difficult because of their heterogeneous nature with respect to specificity (binding to the

---

phase. Nonspecific binding sites blocked with nonspecific protein (e.g., bovine serum albumin or gelatin). (b) Enzyme-labeled antigen incubated in the wells, with or without sample antigen of unknown concentration. (c) The labeled and unlabeled antigens compete for binding to the solid-phase-immobilized antibody. Unbound moieties removed by washing. (d) Substrate for the enzyme added, and following an appropriate period of incubation, the optical density of the colored product is measured. The amount of colored product formed is inversely proportional to the "unknown" antigen concentration, as the higher the free-antigen concentration, the less of the labeled antigen that will bind. A slightly different format of a competitive assay is the *antigen inhibition assay*. In this system, antigen is attached to the solid support and a competition is set up between the free standard or test antigen and the enzyme-labeled antibody. The more free antigen available to bind to the labeled antibody, the less labeled antibody that will bind to the immobilized antigen. Again, an indirect relationship results between the optical density of the product resulting from the enzymatic activity and the antigen concentration.

correct antigen), affinity (the strength of binding due to precision of fit between CDR and antigenic determinant), and classes and subclasses of immunoglobulins present. Different immunoglobulin classes have different valencies, which also affects the avidity (strength of binding). Low-affinity antibodies can still bind strongly to their antigen if they bind more than one epitope at the same time. Monoclonal antibodies, on the other hand, are homogenous with respect to the properties listed above, and immunoassays developed employing them are therefore more reproducible. However, monoclonal antibodies are time-consuming, expensive, and relatively difficult to produce, compared to polyclonals; therefore, their absolute requirement for an assay must be justified. Great care should be taken when screening for monoclonal antibodies for use in immunoassays, to ensure that they do not cross-react with other antigens.

The most frequently used enzymes used for enzyme-linked immunoassays are horseradish peroxidase, alkaline phosphatase, and β-galactosidase. The properties of these enzymes are discussed in Section 4.5.1. Fluorophores, such as fluorescein, have also been used to label antibodies. Fluorescently labeled antibodies are widely used for histochemical applications in conjunction with specialized fluorescent microscopes. Fluorescently labeled antibodies are currently being utilized as the detectable entity in biosensors based on the evanescent wave technique [25]. With the development of microtiter wellplate readers for fluorimeters, fluorescently labeled antibodies and antigens may find a new place in immunoassay technology. No substrates are required as the fluorescence is measured directly. Both radioactive and nonradioactive iodine may also be used as labels in immunoassay systems.

Commercially available 96-well microtiter plates, made of PVC or polystyrene, will bind proteins spontaneously under appropriate conditions and so are a common choice of solid phase for immunoassays. For biosensors, however, the solid support assembly can range from glassy carbon electrodes to silica-based optical fibers to metallic surfaces. Different immobilization chemistries have been developed so that biological molecules can be attached in active form. These will be discussed in the next section.

4.3.6.   Problems Associated with the Use of Antibodies in Biosensors

Antibodies, like most other biological molecules, function effectively only under a certain limited range of physiological conditions. The effects of, for example, pH and temperature on antibody–antigen interactions have been well documented, and assays must be carefully controlled to ensure reproducibility. Analytical samples from different sources must be physiologically similar to reference samples so that one can be assured that the rate of antibody–antigen binding is due to the concentration of the analyte alone.

Immunosensors are based on the detection of the antibody–antigen binding reaction. In order to use an immunosensor several times, the antibody–antigen bond must be broken. Antibodies however, generally have high affinities for their antigens. These high-affinity constants (10^8–10^{12} M^{-1}) confer high sensitivity to immunologic systems. Therefore, breaking the antibody–antigen bond may require drastic conditions that may denature the antibody in some way. Care must be taken in choosing antibodies for reuseable immunosensors. The situation could be somewhat relieved by choosing monoclonal antibodies of low affinity, but the price to be paid here is low sensitivity.

Adsorption as a means of immobilization of antibodies to biosensing surfaces is not acceptable for commercialization because of the problems of leaching. Immobilization of antibodies to different substrate types can involve complex chemistry using hazardous reagents. Antibody orientation must be such that the binding site is available to the antigen. Commercially viable sensors must be able to withstand long-term storage without inactivation of the biological component of the sensor.

Certain immunoassay formats as described in Section 4.4.4 may require some operator skill as additional reagents (e.g., enzyme substrates) may be required in order to detect the immunoreaction. Biosensors for use at the "bedside" by nonscientific people ideally require reagentless systems [12,26]. Biosensors based on mass accumulation measurement may be developed for on-site analysis in the future.

4.3.7. Case Studies on the Use of Antibodies in Biosensors

4.3.7.1. Single-Step Electrochemical Immunoassay Biosensor

Schramm et al. [27] reported the development of an immunosensor with the potential for the determination of several different analytes, which combined the techniques of enzyme-linked immunoassays, electrochemical signal detection, and immunochromatography. The sensor design eliminates the requirement for skilled operators as all necessary reagents are built into the components of the sensor and the timing of the immunoassay is controlled by the immunochromatography mechanism. Such a sensor facilitates on-site analysis.

The sensor design consists of two main parts: an electrochemical sensor and an immunochromatography membrane. The sensor can be used for competitive and noncompetitive immunoassays. Basically, the tip of the chromatographic membrane is immersed in the analyte-containing medium that migrates distally, dissolving the enzyme-conjugated analyte that has been dried onto a spot of the membrane on its way. Antibody, specific for the analyte, is immobilized further along the membrane and is superimposed on the electrodes, a screen-printed working (Pt), auxiliary (Pt), and reference (Ag/AgCl) electrode (Figure 4.12). When the migrating medium comes into contact with the immobilized antibody, the labeled and unlabeled analytes compete with each other for the limited amount of binding sites. The amount of enzyme-labeled analyte bound to the antibody is inversely proportional to the amount of analyte present.

The membrane system reduces the requirement for operator skill as the mem-

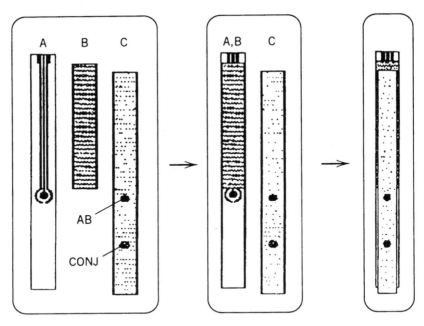

Figure 4.12. Components of the electrochemical immunosensor. It consists of three components (*left*): a chip with the electrodes (A), an insulator (B), and a chromatographic membrane (C). The insulator covers the three leads of the working, auxiliary, and reference electrodes (*middle*). The chromatographic membrane overlays the sensor (*right*). All the components are assembled into a cartridge (not shown). The chromatographic membrane (C) contains an antibody (AB) chemically bound to the membrane and an analyte–enzyme conjugate (CONJ) applied to the membrane so that it migrates with the medium stream.

brane meters a defined volume of specimen medium, serves as a carrier of reagents, and facilitates the separation of bound from unbound analyte at the antibody binding site by capillary flow. Signal detection is achieved by electrochemical oxidation of hydrogen peroxide. Glucose oxidase, the enzyme label of the conjugate, catalyzes the formation of H_2O_2 from glucose and oxygen. The substrate for the enzyme, glucose, can be preapplied to the membrane or placed at the tip of the membrane so that it migrates with the analyte-containing medium. H_2O_2 formed before the immunoreaction takes place is conveniently removed by the continuous migration of the medium.

Antibodies can be produced against a large number of analytes and applied to this system. The system, however, must be optimized for different analytes. The retention of molecules transported by capillary flow can vary depending on combinations of various interactions between solid and liquid phases.

4.3.7.2. *Evanescent Wave Fiber-Optic Immunosensor*

Ligler et al. [25] reported the development of an evanescent wave fiber-optic immunosensor that facilitated the analysis of environmental and clinical samples for hazardous materials. By using long fibers, contamination of the operator and optical equipment by potentially hazardous materials is avoided. When a light ray is totally internally reflected in an optical fiber within a certain range of angles, a fraction of the radiation extends a short distance from the light-propagating medium into the medium of lower refractive index. This nonpropagating electromagnetic radiation is called the *evanescent wave*. An optical fiber typically consists of a transparent silica-based fiber core surrounded by plastic cladding of lower refractive index. If the cladding is removed, however, we have a situation where molecular species can interact with the evanescent wave radiation as the sample solution now acts as the cladding or lower-refractive-index medium.

The nature of the evanescent wave is such that it only interacts with molecular species within its penetration depth. Antibodies immobilised onto the exposed core of an optical fiber bind antigens in solution, concentrating them within the evanescent sensing zone (Figure 4.13). Detection is by means of fluorescent labels. On absorption of radiation, for example, in the form of a photon, a molecule will attain an excited energy state. The extra energy gained by the molecule is subse-

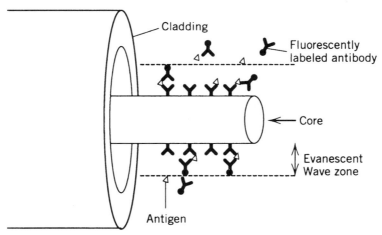

Figure 4.13. The nature of the evanescent wave is such that it interacts only with molecular species that lie within its penetration depth. Antibodies are immobilized on the exposed core of a silica-based fiber. Antigens and fluorescently labeled second antibodies form a "sandwich" immunocomplex with the immobilized antibody. This immunocomplex lies within the evanescent zone, and is detected. Unbound antigen and fluorescently labeled antibody remain in the bulk solution and thus are not detected by the evanescent wave; therefore, no separation step is required.

quently released. This process generally occurs by collision with other molecules or in a chemical reaction. In cases where the excited state of the molecule is sufficiently stable, however, the surplus energy is released by means of emission of radiation. This phenomenon is known as *fluorescence.* The emitted radiation is at a longer wavelength than that of the excitation energy. By exciting the molecules in a given solution and measuring the intensity, if any, of the fluorescence produced, the amount of fluorescent material in the solution may be estimated. In biosensors of this type, the evanescent wave interacts with the fluorophores and the resulting fluorescence is coupled back into the fiber and detected at the distal or proximal end. Antibody–antigen interactions are detected in real time as the fluorescent label accumulates onto the fiber core simultaneously with the antigen of interest. Antigen concentration can be related to either the rate of fluorophore accumulation or the intensity of fluorescence at a certain timepoint.

Ligler et al. [25] immobilised antibotulinum toxin A to the core of 200-μm-diameter, 1-m-long tapered optical fibers. Tapering the fibers improves the amount of fluorescence that can be coupled back into the fibers and therefore increases the signal. The affinity-purified polyclonal antibodies and a murine monoclonal antibody against botulinum A were labeled with the fluorophore, tetramethylrhodamine-5-isothiocyanate (TRITC). One-step and two-step noncompetitive assays were performed using the monoclonal and polyclonal fluorescently labeled antibodies, respectively. Detection limits of 1 ng/mL toxin A were established. Stability studies showed that the antibody-coated fibers were stable for up to 19 months when stored frozen in PBS or lyophilized.

4.4. IMMOBILIZATION OF BIOMOLECULES FOR BIOSENSORS

Fundamental to biosensors is the biosensing surface made up of enzymes, antibodies, antigens, microorganisms, mammalian cells, tissues, or receptors immobilized on to a solid surface. A number of established immobilization procedures are now used. These include

1. Physical adsorption on to a solid surface
2. Use of crosslinking reagents
3. Entrapment using a gel or polymer
4. Use of membranes to retain the biomolecule close to the electrode surface
5. Covalent attachment
6. Exploitation of biomolecular interactions

The choice of immobilization method used depends on the biocomponent to be immobilized, the nature of the solid surface, and the transducing mechanism.

Whatever immobilization method is used, it must take into account the following factors:

1. The activity of the biomolecule must be preserved, and the specificity must not be reduced.
2. If possible, its stability should be maintained or preferably increased. This is important in relation to the stability of the sensor system for use, reuse, and storage.
3. Steps should be taken to ensure that no nonspecific binding can occur as a consequence of the immobilization procedures used.
4. The methods used should ideally be easy to perform and highly reproducible. Otherwise, large-scale production of the sensor will not be feasible.
5. Extreme conditions (e.g., pH, temperature, salt concentration) should be avoided if possible, but it should be noted that many biomolecules, given the correct conditions, can be subjected to relatively high temperatures for short periods. By using biomolecules from thermophilic organisms, chemical and genetic modification, it should be feasible to produce designer molecules capable of being reactive under defined conditions. This area is now under extensive investigation.

The established immobilization methods are described in further detail here:

1. *Physical Adsorption.* Immobilization via physical attraction or adsorption is not a reproducible and reliable method of biomolecule attachment to sensing surfaces because of the problems associated with leaching during long-term storage. Plastic, glass, and cellulose have been known to adsorb proteins via binding forces such as hydrogen bonds, van der Waals forces, salt linkages, and hydrophobic interactions. Such forces are not very stable and can be easily disrupted by changes in pH, temperature, and ionic strength. Excess protein can also form multiple layers during adsorption. These layers are not very stable and easily desorb. The obvious advantage of adsorption as a means of immobilization is its simplicity and gentleness. However, in the case of antigens, care should be taken to ensure that the antigenic binding sites are not denatured in any way and that they are oriented in such a way as to allow recognition.

2. *Use of Crosslinking Reagents.* Stabilization of adsorbed proteins can be achieved by using bifunctional crosslinking reagents such as glutaraldehyde. The proteins are crosslinked to each other or inert proteins such as bovine serum albumin can be mixed with the desired protein prior to crosslinking. This adds greater stability to the immobilized protein, although inevitably, some inactivation does occur as the crosslinking chemical may interact with the active site, in the case of an enzyme. With respect to immunosensors, the antibody binding site may be

blocked or incorrectly oriented so as not to favor antibody–antigen binding. Membranes can be cast on electrode surfaces using this method.

3. *Entrapment.* Immobilization of biomolecules by physical entrapment in gel matrices such as polyacrylamide or gelatin, for example, is a mild method of immobilization. Basically, the polymer is allowed to crosslink in the presence of the substance to be immobilized. The porosity or degree of crosslinking can be controlled. Evidence has shown that polyacrylamide gels give the best retention of enzymatic activity. Care should be taken to ensure that the crosslinks retain the molecule and that leaching does not occur. Microorganisms are often immobilized by physical entrapment.

4. *Use of Membrane to Retain the Biomolecule Close to the Transducer Surface.* Membranes of various porosities can be used to retain molecules close to transducer surfaces without the need for actual immobilization. This gentle method of retention can, however, lead to problems such as diffusional resistance. "Selective" membranes can also be employed in conjunction with potentiometric electrodes. Ion-selective membranes, such as an NH_4^+ selective membrane, has been employed in conjunction with the enzyme, urease, for the measurement of urea [28].

5. *Covalent Attachment.* Chemical coupling of biomolecules may provide stable biosensing surfaces that are resistant to wide ranges of pH, temperature, and ions. Covalent binding can result in some loss in bioactivity. Three types of supports have been used: inorganics, natural polymers, and synthetic polymers. The binding process must occur under conditions that do not denature the biomolecule. Often, the carrier must be activated in some way (e.g., silanization), and the introduced functional groups are then utilized for chemical coupling either directly to the biomolecule or via a crosslinking reagent such as glutaraldehyde.

6. *Exploitation of Biological Interactions.* This approach has most potential due to the absence of many inherent limitations of other immobilization strategies. Immobilization of antibodies has been achieved based on the specific association of Fc receptors with protein A [29] and protein G [30]. A recombinant protein A/G has been developed which combines the IgG-binding profiles of the parent molecules [31]. More recently the ability of avidin and biotin to form a tight, highly associated noncovalent complex has been exploited [32]. Deposition of avidin on electrodes has been achieved electrochemically [33,34], by stepwise adsorption [34] and covalent attachment [35]. Biospecific electrode biosensors used in the determination of trace concentrations of analytes are described [36–39]. An extrapolation of this approach is the production of biotinylated antibodies, directed against an analyte of interest, which can be immobilised on an avidin coated surface [40,41]. Optimal antigen binding can be achieved with this approach as the variable regions are not altered and the antibody is positioned in a favorable orientation. The area of enzyme, antibody, and receptor immobilization for biosensors is discussed in excellent reviews by Guilbault et al. [42] and Taylor [43].

4.5. MODIFIED AND HYBRID BIOMOLECULES: USE AND POTENTIAL IN BIOSENSORS

Modified and hybrid biomolecules have been extensively used as the biological component in biosensors. The high specificity and ease of production of antibodies—particularly with the advent of monoclonal antibody technology—has made them the main target for modification. In general, two approaches have been used in this process. The first of these—involving labeling of the antibody—has been discussed in Section 4.3.5. In the conjugation of label and antibody, some basic considerations apply. The label to be used should possess chemical groups that enable it to be covalently linked to the antibody molecule. It is, of course, necessary that conjugation of the label not disrupt the immunologic properties of the antibody. Preservation of the molecule's antigenic nature may also be a consideration. Stability of the conjugate is another important factor, and should be similar, under usual storage conditions to that of the unconjugated antibody. For practical purposes, the conjugation procedure to be used should be as rapid and straightforward as possible, and use as inexpensive and inoffensive reagents as practicable. It should be noted that biosensors that employ labels can operate on either the sandwich or competitive assay approach. In a sandwich assay, sequential binding of analyte and labeled antibody is involved. The competitive format, however, may make use of either labeled antibody or labeled analyte. The labeling methods discussed below may therefore also be applied to labeling of the analyte.

The second type of modification of biomolecules entails the production of hybrid immunoglobulin molecules such as bifunctional, single-domain, and catalytic antibodies. In these molecules, the structure of the antibody itself is changed, either to enhance the specificity or to incorporate a signal-generating component. Both biological and chemical means may be used in their production.

4.5.1. Enzyme-Labeled Antibodies

Of all the possible labels for biomolecules in sensor systems, enzymes are probably the most commonly used and widely studied. By combining the specificity of antibodies with the degree of signal amplification afforded by the catalytic effects of an enzyme, a high level of both selectivity and sensitivity can be obtained. An outline of the advantages of enzymes and some examples of their use in biosensors is given in Tables 4.1 and 4.2.

One major advantage of the use of enzyme labels is the ease with which they can be conjugated to antibodies, with very little resultant loss of biological activity. Several different approaches are available for the conjugation of enzymes to antibodies.

The conjugation of glycoproteins, such as peroxidase, is best carried out using the periodate method. In this procedure, (first described in 1974 by Nakane and

Kawaoi) carbohydrate residues on the enzyme are oxidised by periodate to produce aldehyde groups, which react with free-amino groups on the antibody. The procedure is an efficient and relatively straightforward one. The enzyme is dissolved in water, freshly prepared $NaIO_4$ is added, and the mixture is stirred for a short time at room temperature. Following a dialysis step to minimise the risk of autoconjugation, the antibody is added in carbonate buffer, and the reaction mixture is again stirred. Sodium borohydride is then added to stabilize the conjugate by preventing hydrolysis.

Crosslinking, using either heterobifunctional or homobifunctional reagents, is an alternative method of conjugating enzymes and antibodies. These reagents are molecules that contain two functional groups, which may be the same (homo-) or different (hetero-). There are two approaches to this type of conjugation: the one-step procedure, in which enzyme, crosslinker, and antibody are added together; and the two-step method, where the enzyme and crosslinker are reacted first, and antibody is then added, following removal of excess reagent.

One of the most widely used crosslinking reagents is the homobifunctional crosslinker glutaraldehyde. Again, the conjugation procedure is fairly simple. Enzyme and antibody solutions are first dialyzed thoroughly against phosphate buffer to remove interfering substances. Glutaraldehyde solution is then added dropwise. The reaction mixture is incubated at room temperature for several hours, and any remaining reactive sites are blocked by the addition of an excess of L-lysine.

Other bifunctional reagents are also available. N,N'-o-Phenyldiamaleimide (OPDM) is suitable for use in the conjugation of proteins that contain thiol groups, or into which thiol groups can be introduced. 5-Acetylmercaptosuccinic anhydride (AMSA) and methyl-4-mercaptobutyrimide (MMBI) are both convenient reagents for the introduction of thiol groups into proteins. Another crosslinking reagent that may be used is bis(succinic acid) N-hydroxysuccinimide ester.

Many biosensors have been demonstrated that use a redox mediator coupled to an enzyme-labeled antibody. A sandwich assay has been described for the detection of human chorionic gonadotrophin (hCG). The method uses two monoclonal antibodies to different sites on the hCG molecule. In the first step, glucose oxidase–labeled anti-hCG is reacted with HCG in the sample. Antibody immobilised on the electrode is then used to capture the labeled antibody–antigen complex. The activity of the bound label is then measured by cyclic voltammetry, using a mediator [44].

Potentiometric methods may also be applied to measure the products of enzyme labels. One example of this is the detection of hepatitis-B surface antigen (hBSAg) using horseradish peroxidase–labeled anti-hBSAg. Unlabeled antibody is first immobilized on the surface of an iodine-selective electrode, and placed in the sample solution to extract hBSAg. The electrode is then dipped into a solution of HRP-labeled antibody, to form a "sandwich." By placing the electrode into a peroxide iodide solution, one may determine the amount of labeled antibody present by following the reaction [45]:

$$H_2O_2 + 2I^- + 2H^+ \xrightarrow{\text{HRP}} I_2 + 2H_2O \qquad (4.23)$$

As mentioned previously, sensors based on the Clark oxygen electrode have been developed. One example is that of a sensor for the detection of the tumor marker alphafetoprotein (AFP). Anti-AFP is first immobilized on the sensor surface. In a competitive system, free and catalase-labeled AFP then compete for binding sites on the surface, and the degree of catalase activity is determined by amperometric measurement, following the addition of hydrogen peroxide [46].

4.5.2. Fluorescently Labeled Antibodies

The use of fluorescently labeled antibodies in evanescent wave biosensors has been discussed previously (see Section 4.3.7.2.). In some cases fluorescence may be detected at a concentration as low as single molecules. For highly sensitive levels of fluorescent detection, and for applications in optical fiber technology, lasers are generally used as the source of excitation light. Of these, the argon ion laser is the most commonly used. Using this type of excitation does, however, place certain limitations on the fluorophore that is employed. The main requirement is the need to choose a molecule that absorbs at the discrete wavelengths produced by the laser.

Another problem often encountered in the area of highly sensitive fluorescent detection is that of *photobleaching,* a process in which the fluorescent dye is destroyed by the excitation light, and new, usually nonfluorescent, products are formed. The rate of photobleaching of a fluorophore will limit the number of times that a molecule can be excited. Protective agents, such as *p*-phenylenediamine and propyl gallate, may be used to diminish the effects of photobleaching. It should be noted that the emission and excitation spectra and lifetimes of fluorescent dyes may be affected by the type of immobilization used to link the labeled biomolecule to the sensor surface. The transient nature of the signal and the presence in most biological fluids of intrinsically fluorescent molecules that will increase the background noise must also be considered as drawbacks. Fluorophores are also acutely sensitive to their environment. Factors such as pH, ionic strength, type of solvent, viscosity, temperature, and additional chemical groups can all have an effect on the observed fluorescence. These factors mean that the extent of chemical modification that a fluorescent molecule may undergo is severely curtailed. Modification of a dye, in order to introduce a functional group that will react with a site on the antibody, can often result in severe or total loss of fluorescence. Nonetheless, in recent years many fluorescent dyes with a wide range of functional groups for incorporation into biomolecules have become available.

A variety of different groups exist on antibody molecules for the coupling of fluorescent dyes. In addition to the terminal amino and carboxyl groups, many free amino groups are present in lysine side chains and aspartic and glutamic acid side chains will contain free-carboxyl groups. There are also guanido, imino, thiol, and phenol groups. Notwithstanding this multiplicity of available sites, however, the vast majority of antibody–dye conjugations are carried out either at the thiol group of cysteine or the ε-group of lysine.

Fluorescein isothiocynate (FITC) is among the most commonly used dyes for antibody labeling. Fluorescein has several advantages as a fluorophore. These include a good chemical stability, high absorbance, and the fact that its reactive derivatives are water-soluble above pH 6. Although the most widely used derivative is FITC, many others have been described. Examples of these are amine-reactive dichlorotriazines, disulfides, and succinimidyl esters.

Fluorescein absorbs at a wavelength very close to the argon ion laser line of 488 nm, and emits with a high intensity. It does, however, possess a few inherent disadvantages. The phenolic group of the molecule has a pK_a of about pH 6.4. Acidification below pH 7 quenches the absorbance of fluorescein considerably. A significant degree of photobleaching has also been observed with this dye. In addition to this, emission by fluorescein derivatives in serum can be interfered with by the presence of bilirubin.

Rhodamines are more stable to illumination than are the fluorescein derivatives and are also less sensitive to pH. However, they have a lower fluorescent intensity. Several rhodamine derivatives are available. Tetramethylrhodamine–isothiocyanate (TRITC) is very suitable for excitation by the 546-nm line of the mercury arc lamp. Rhodamine B has longer wavelengths of absorption and emission than does TRITC. Sulfarhodamine B (also called *Lissamine rhodamine B*) is a derivative of rhodamine B in which the sulfonic acid groups have been converted to sulphonyl chlorides. Sulfonamide groups formed by this compound during conjugation are more stable than the bonds formed by isothiocyanate derivatives. Introduction of chloride groups into the compound also involves only minimal disruption of the structure of the dye, so that the amount of fluorescence produced is not greatly affected. Texas Red is a derivative of sulforhodamine 101. It has the longest emission wavelength of any of the dyes in general use, and a higher fluorescent intensity than either TRITC or the rhodamine B derivatives. It has a tendency, however, to cause the precipitation and aggregation of proteins to which it is conjugated. In particular, it causes the precipitation of antibodies. As a result, it is usually used as an indirect labeling agent—for example, as its protein A or avidin conjugate. Technically, the conjugation of antibodies to the isothiocyanate derivatives of fluorescein or rhodamine is a relatively simple matter. Antibody and dye are mixed, and a high degree of labeling can be achieved within 1 or 2 h. It has been demonstrated that a fluorescein:antibody ratio of 1:4 is optimal for labeled antibodies. Care should be taken when using the derivatives, as they are prone to hydrolysis even in air. In the course of the conjugation procedure, a competing hydrolysis reaction occurs. Conjugation is favored at high antibody concentrations, but at lower concentrations the significance of the competing hydrolysis decreases. It should be noted that high levels of conjugation with TRITC and Texas Red actually give rise to a decreased fluorescence, when compared to that observed with less heavily conjugated antibody molecules. This is due to self-quenching of the fluorophores, and should be borne in mind when choosing antibody:dye ratios.

The phycobiliproteins are a group of highly fluorescent algal pigments that can be used for labeling antibodies. More elaborate conjugation procedures are necessary to link these compounds to biomolecules. In nature, they are responsible for light gathering and energy transfer to chlorophyll. As such, they are optimized for fluorescent intensity, and are used to enhance sensitivity. In practical applications, the enhancement with regard to fluorescein has been reported to be in the range of a fivefold increase in levels of fluorescence. A broad spectrum of excitation wavelengths is also observed, due to the fact that several different types of pigment are present in each protein. Only the emission from the pigment of longest wavelength is seen. One inherent disadvantage of these molecules is that, being proteins, they do not possess the chemical reactivity of other dyes, and are therefore less easy to conjugate. Labeling with these fluorophores is generally a three-step method, involving the addition of thiol groups to the phycobiliproteins, and thiol-reactive maleimide to the antibodies.

7-Amino-4-methylcoumarin-3-acetic acid succinimidyl ester (AMCA) is a fluorescent dye with an absorbance in the visible range of the spectrum and a bright-blue fluorescence at 450 nm. This compound is available commercially in a conjugate with NHS, sulfo-NHS, hydrazide, HPDP, or streptavidin. It has an optimal excitation wavelength of 351/304 nm with a UV argon laser.

A variety of biosensors using fluorescent-labelled biomolecules as the biological sensing component have been reported, based on both competitive and sandwich approach. The method of Ligler et al. [25] for the detection of botulinum toxin A has been described in detail in Section 4.3.7.2. FITC has been used as a fluorescent label for rabbit IgG, in a competitive assay for human IgG [47]. In this assay, a protein A monolayer on the sensor was exposed to both human IgG and FITC-labeled rabbit IgG. The observed fluorescence that resulted was inversely proportional to the concentration of the analyte. β-Phycoerythrin has been used to fluorescently label lidocaine in a competitive fiber-optic biosensor system [48]. Texas Red has been used, along with fluorescein-labeled antigen, in a biosensor that operates using an energy-transfer approach. In this assay, fluorescein-labeled and free antigen compete for a limited number of sites on a Texas Red–labelled antibody surface.

4.5.3.   Iodine-Labeled Antibodies

The ability of a trace amount of iodide to catalyze the reduction by arsenic(III) of cerium(IV) can be used to measure the amount of iodide present in a reaction mixture. The reaction, first described by Sandall and Kolthoff in 1934 [49], is as follows:

$$Ce^{4+} + As^{3+} \rightarrow Ce^{3+} + As^{5+} \tag{4.24}$$

By monitoring the progress of this reaction, it is possible to determine the concentration of iodine present. Thus, "cold" (nonradioactive) iodine can be used as a label in biosensor systems.

Substitution methods of iodination depend on the electrophilic replacement of the hydrogen atom in the tyrosyl or histidyl component of the antibody with iodine. The aromatic ring of tyrosine and the imidazole ring of histidine are open to electrophilic attack at pH 7 and 9, respectively. The iodous ion (I^+) acts as the electrophilic agent. To completely eliminate contact between the antibody and the powerful chemical agents used in substitution procedures, a conjugation method may be used. Conjugation methods are also the only way of iodinating antibodies that do not have tyrosine residues. A summary of the various different methods of iodination is given in Table 4.6.

The choice of iodination method will depend greatly on the relative severity of the procedures. Where oxidation methods are employed, the yield of labeled antibody may be quite low, due to exposure to strong chemical agents. In addition, the iodide ion is roughly the same size as the phenolic ring of tyrosine, which is the primary site of incorporation in substitution methods. Many tyrosine residues are closely associated with biological activity, which may lead to high levels of inac-

Table 4.6. Methods of Iodinating Proteins

Method	Comments	Reference
Substitution		
Chemical		
Chloramine-T	Cheap, simple, short reaction time	50
Iodobeads	Gentler method than Chloramine-T	51
Iodogen	Less likely to denature protein	52
Iodine monochloride (ICl)	Stable reagents, simple; level of iodination can be controlled	53
Enzymatic		
Enzymobeads	No exposure to oxidizing agents	54
Electrolytic iodination	Complex, rarely used	55
Conjugation		
Bolton–Hunter reagent	Less conformational change to antibody caused	56
Aniline		57
Methylparahydroxybenzidimate	More selective for amino groups than Bolton–Hunter reagent	58

tivation if such an active site is blocked with one or two iodine ions. However, conjugation methods are gentler, and thus have slower rates of reaction, which may again result in relatively low yields. As^{5+} is not an electroactive species, so the use of the Sandell–Kolthoff reaction in biosensors relies on measuring the decrease in concentration of As^{3+}, by means of an ion-selective electrode.

4.5.4. Antibody Fragments

The production and function of the various antibody fragments have been outlined in Section 4.3.3. These fragments may be utilized for the construction of biosensors. In particular, F(ab')$_2$ and Fab fragments may be used as the immobilized component on the sensor surface. A sensor has been described that uses immobilized F(ab')$_2$ fragments to facilitate reversible fixation of antigen-antibody complexes [59]. In addition, Fab fragments are of use in the manufacture of bifunctional antibodies.

4.5.5. Bifunctional Antibodies

Among the most interesting of the newly emerging hybrid biomolecules are the bifunctional antibodies. These immunoglobulins are bivalent antigen-binding molecules, in which the two arms of the antibody are specific for two different antigens. The use of a bifunctional antibody can obviate the need for labeling. The chemical means used to conjugate labels such as iodine molecules, enzymes, and fluorescent substances to antibodies may damage the biological efficacy of the immunologic component. Both the activity and the stability of the antibody can be affected by the conjugation process. If the label is bound by one of the two arms of the antibody, however, there is no need for chemical modification of the structure. Also, as one half of the bifunctional antibody can be raised to a specific epitope of the label, inhibition of enzyme labels can be avoided.

Bifunctional antibodies were first developed for use in cancer chemotherapy by targeting tumor cells and providing close contact with the chemotherapeutic agent [60]. Both biological and chemical methods for their production have been developed. Biological production is based on conventional monoclonal antibody technology. By fusing two hybridomas of different specificities, bifunctional monoclonal antibodies can be produced [61]. These are known as *quadromas,* as they are derived from four parental cells. A variation on this method consists of fusing a hybridoma specific for one antigen with splenocytes from an animal immunized with a second antigen [62]. The resultant somatic hybrid is referred to as a *trioma,* as it is derived from three parental cells. The production of quadromas has proved to be the more extensively-used method, as it is easier to work with a population of cells that are producing antibodies of known affinity, rather than an heterogenous mixture of spleen cells producing uncharacterized antibodies. Although both of these biological methods have been used with a high degree of success in the production of bifunctional antibodies, the amount of time needed to perform them is a drawback. This factor, along with the unavoidable production of a mixture of both bifunctional and parental antibodies in biological systems, has lead to the development of chemical methods of manufacture. Basically, these consist of cleaving two antibodies and joining Fab fragments of differing specificities.

Pepsin hydrolysis is used to produce $F(ab')_2$ fragments, which are then reduced to $F(ab')$ in the presence of the dithiol complexing agent sodium arsenide, which prevents the formation of intrachain disulfide bonds between the three thiol groups of mouse IgG. The thiols are then activated as thionitrobenzoate (TNB) derivatives by the action of Ellman's reagent. Reconversion of one of the $F(ab')$ TNB derivatives to $F(ab')$-thiol is then carried out by reduction with 2-mercaptoethanol, and this is then reacted with an equimolar amount of the remaining $F(ab')$-TNB, to produce a bifunctional antibody. The biospecificity of the product can be tested using a double-antigen ELISA. Yields of up to 70% are possible with this method.

For the construction of bifunctional antibodies consisting of mouse and rabbit Fab fragments, a separate approach has been devised. Reduction of $F(ab\gamma')_2$ fragments of both antibodies gives high yields of $F(ab\gamma')$—SH fragments. O-Phenyldiamaleimide, a crosslinking reagent, is then used to maleimidate all of the —SH groups on one of the antibodies. The $F(ab\gamma')$-SH and $F(ab\gamma')$-Mal fragments are then reacted together in conditions that favor —SH and maleimide combination, at low pH, in the presence of EDTA. Again, yields of up to 70% of the theoretical maximum may be obtained using this method.

A different approach to the production of bifunctional antibodies is the development of heteroconjugates, consisting of two separate monoclonal antibodies that have been chemically crosslinked. One way of doing this is to react both parent antibodies with N-succinimidyl 3-(2-pyridyldithio) proprionate (SPDP), forming pyridyl disulphide antibodies. One of these is then further reduced with dithiothreitol. The two are then reacted together. Alternatively, one of the parent molecules may be derivatised with 2-iminothiolane, and reacted with the other, which has been previously treated with SPDP. Resultant heteroconjugates can be separated from homoconjugates and derivatised unconjugated antibodies, by virtue of their different isoelectric points.

Recently, the production of small bivalent and bispecific antibody fragments has been described [63]. These molecules, called *diabodies,* consist of a heavy-chain variable domain (V_H) linked to a light-chain variable domain (V_L) on a single polypeptide chain. By using a spacer arm, the C terminus of one domain may be joined to the N terminus of the other. If this spacer arm is kept short, two domains on the one chain cannot associate. If V_H and V_L domains of two different antibodies, A and B, are linked to form the chains $V_H A$–$V_L B$ and $V_L A$–$V_H B$, which are coexpressed in the same cell, then the chains should associate to form a molecule with two antigen-binding sites. These small bispecific antibody fragments assemble in vivo, and can be extracted directly from the cell culture supernatant. The diabodies are roughly equivalent in size to a Fab fragment.

Bifunctional antibodies have remarkable potential for applications in the construction of biosensors. In addition to removing the need to label antibodies chemically, these molecules can also facilitate more efficient immobilisation of the labels. Initial immobilization of antigen on sensor surfaces can be followed by binding of bifunctional antibodies with affinity for an enzyme or fluorescent molecule, at an epitope on that molecule that doesn't quench, or inhibit the active site—thus diminishing the concomitant loss of activity often seen with chemical immobilization methods.

4.5.6. Single-Domain Antibodies

It is possible to produce isolated V_H domains that have a relatively high affinity for antigen, obviating the necessity to link them to V_L domains. These antigen-binding domains are called *single-domain antibodies* (dAb's). The spleen genomic DNA of immunized mice can be used to clone V_H genes using the polymerase chain reaction. These genes can then be expressed in *E. coli*.

This process can be carried out in a matter of days, with no need to use tissue culture. The V_H domains have been shown to exhibit specificities for antigen of a similar degree to those shown by monoclonal antibodies for their protein antigens. It is envisioned that these dAb's will become valuable building blocks for the construction of high-affinity antibodies for use in human therapy, and catalytic antibodies. Their potential applications in the field of biosensors are wide and exciting.

4.5.7. Abzymes

Although there may appear at first glance to be significant differences between the structure and function of enzymes and those of antibodies, the potential of a hybrid molecule, combining the specificity of an antibody with the catalytic action of an enzyme, has resulted in the production of a range of catalytic antibodies, or *abzymes*. The potential applications of such a molecule are enormous. In medicine, an antibody with the catalytic effect of a restriction enzyme could be used in the treatment of cancerous cells and pathogenic viruses. Industrially, these molecules could be used as synthetic agents in chemical and pharmaceutical processes. In the area of biosensors, the applications of an abzyme are almost endless.

Catalytic antibodies possess several advantages over enzymes. With abzymes, the protein structure is largely conserved, except for the variable antigen-binding regions. This is in contrast to the wide variety in enzyme structure. In addition, the stability of abzymes is often greater than that of enzymes—so that the construction of sensors with longer lifetimes is a possibility.

Both enzymes and antibodies have the same basic mechanism for recognizing their target molecules—via a specialized binding site. In the case of antibodies, recognition is achieved by a variety of weak bonding forces, such as van der Waals and electrostatic interactions, and hydrogen bonding. The facility to make a hybrid molecule came about as a result of emergent monoclonal antibody technology, and an understanding of the importance of transition states to the catalytic effects of enzymes.

In the course of a chemical reaction in which one molecule is transformed into another, the energy profile of the reaction shows the initial molecule and the final product (both of which are stable structures) lying in deep thermodynamic "wells." In order for the reaction to occur, the atoms of the molecule must first gain energy to move out of their well, and then lose energy to fall into the product well. At the highest point in this energy profile, an unstable "transition state" of the molecule is present (see Figure 4.3). The difference between this energy level, and that of the initial molecule is termed the *activation energy,* and is the amount of energy needed to drive the reaction to completion. The catalytic effect of enzymes is due, to a large degree, to the fact that they preferentially bind to the transition state, rather than the ground state, of their substrates. Hence, less energy is needed to reach the transition state, and the rate of reaction is increased.

Antibodies, on the other hand, bind to the ground state of their target molecules. By producing an antibody specific for the transition state, then, the same catalytic effect should be observed. However, as the transition state is by definition unstable, and exists only briefly, it is necessary to use an analog of the structure to raise antibodies. A transition-state analog (TSA) is a molecule with the same size, shape, and charge as the transition state, which resembles it so closely that it will act as a powerful inhibitor of the enzyme. By using TSAs as antigens in the production of monoclonal antisera, many abzymes have been produced, with catalytic activities in the range of 10^3–10^4-fold.

One important property of these abzymes is their stereospecificity. It is possible to carry out the hydrolysis of one chiral form of a molecule in an enantiomeric mixture, with a selectivity of up to 98%. Most of the work done in the field to date has concentrated on the production of abzymes for acyl-transfer reactions, but the use of other TSAs should enable catalysis of various different reactions, such as metallation of mesoporphyrin, hydrolysis of glycosidic bonds, and hydrolysis of phosphodiester bonds.

Increased affinity for the transition state of the substrate is not the only mechanism used in enzymatic catalysis. For reactions of higher activation energy, the side chains in the binding pocket of the enzyme are directly involved. This has the effect of breaking the activation energy of the reaction down into a series of steps, with transition states of lower energy. One example of this process is the serine 195 of chymotrypsin, which undergoes acylation by the substrate. If such amino acid side chains could be introduced into the binding site of an antibody, catalysis could be greatly enhanced. That section of the antibody that is involved in antigen recognition has the highest degree of variability in its amino acid composition. In addition, it is known that complementary structural features are induced in the binding site by antigens. For example, charged groups on the antigen are surrounded by oppositely charged side chains on the antibody, and hydrophobic groups are presented with a nonpolar environment. Thus, by careful antigen design, it is possible to introduce the required amino acids into the binding sites of an antibody. The huge amount of variation in the binding site has the potential for the introduction of a number of different catalytic groups into the same abzyme.

Another factor of importance in the mechanism of enzymic catalysis is that of proximity effects. It is possible to use antibodies as "entropy traps" to lower the entropic barriers of reaction. The binding energy of the antibody is used to negate the rotational and translational motion for reaction. Conformations of the molecule that are necessary for reaction are bound preferentially by the abzymes, giving greatly increased rates of reaction when compared to uncatalyzed systems.

In addition to the more conventional methods of raising monoclonal antibodies to transition-state analogs, both chemical and molecular biological methods are available for the introduction of catalytic groups into antibodies. In this fashion, the antibody can be designed in vitro, with a specific application in mind.

One method that can be used to design the affinity of antibodies is that of site-directed mutagenesis. In order to carry out this process, the three-dimensional structure of the antigen-combining site on the antibody must be known. In the absence of a detailed three-dimensional structure, chemical modification may be used to identify residues near to the binding site. Molecular modeling and epitope mapping may be used to elucidate the structure. Light- and heavy-chain antibody cDNA clones are then inserted into a mutagenesis vector, and an oligonucleotide primer is used to perform the site-directed mutagenesis.

Recently, bacterial expression systems have been developed that greatly facilitate the production of large amounts of abzymes. Protein engineering may also be used to combine the binding sites of the antibody with other proteins. Chemical semisynthesis may also be used to introduce catalytic groups into the combining sites of antibodies. One advantage of chemical modification is that it allows for the introduction of synthetic groups, such as redox active metals, and a wide range of acids, bases, and nucleophiles. Another advantage is that no detailed knowledge of the three-dimensional structure is needed. The procedure is a two-step one. Antibody to a specific antigen is first produced. Then an affinity labeling agent is produced, which contains a cleavable linkage between antigen and affinity-labeling group. The antigen confers the specificity required to direct the labeling group to the amino group of interest. Once the antigen has bound to the antibody, the linkage between labeling agent and antigen is broken. Generally, the labile linkage used is a thiol group, and cleavage results in this reactive thiol group remaining adjacent to the antigen-combining region of the antibody. It is then possible to derivatize this group with a wide range of synthetic catalytic groups. For example, derivatization with metal complexes of porphyrin or flavins gives the potential for redox catalysis. Enzymatic catalysis is greatly enhanced by the action of cofactors. These are nonproteinaceous catalytic auxiliaries, and include flavins, haem, and metal ions. The incorporation of these groups into antibodies can also enhance the range of their catalytic activity. In addition, a variety of cofactors not available for use by enzymes can be introduced. Incorporation of cofactors into abzymes can be achieved in various ways. First, an analog of both cofactor and substrate can be used to produce antibodies to the functional regions of both moieties in a single immunization. Cofactors can also be introduced using the method of chemical semisynthesis, as described previously. Another way of introducing cofactors into antibodies is by exploiting the fact that antibodies possess two chains. One of the chains can be used to bring the cofactor into the binding pocket of the antibody.

In the field of biosensors, abzymes have enormous potential. Compared to the vast repertoire of antibodies produced by the immune system, only a handful of enzymes exist. Catalytic antibody technology provides the opportunity to manufacture an almost endless array of new "designer enzymes," thus removing some of the major limitations of enzyme-based sensors—namely, that they are confined to reactions that occur in nature, and that it is often difficult to purify the desired enzyme. Although still a relatively new field, several prototypes have been developed to assess the effectiveness of abzymes in biosensor systems. One such sensor [64] uses a modified pH electrode to monitor the buildup of acidic products in the course of an abzyme-catalyzed phenyl acetate hydrolysis reaction.

4.6. CELL-BASED SENSORS

Enzymes have undoubtedly been the most commonly employed biological component of biosensors to date, due to the diversity of biochemical reactions they catalyze. There are potentially thousands of organic compounds that can be analyzed on the basis of their involvement in enzymatic reactions that produce metabolites detectable by various transducing mechanisms. However, enzymes are proteins and must be handled carefully in order to prevent denaturation resulting in reduction or loss of enzymatic activity. Prior to the late 1970s, enzymes were purified from their source (microorganism, plant tissue, or mammalian tissue) for use as the sensing element in biosensors. Although purified enzymes offer potentially better selectivity than do crude enzyme preparations—in that interference from other enzymes is significantly reduced or eliminated—enzymes purified from their natural sources have many limitations. Enzymes are most active in their "naturally optimized" environment. The interfacing of biologically intact sensing structures to electrode surfaces is, as yet, an underexploited area of biosensor technology. The advantages of using whole cells in place of isolated enzymes include:

1. Whole tissues or microbes are a source of a large quantity of undamaged enzyme.
2. Enzymatic pathways inherent in tissue cells may be difficult or impossible to reproduce outside the cell.
3. Enzymatic reactions are already optimized in intact tissues.
4. All cofactors, substrates, and so on for enzymatic reactions are available in whole tissues.
5. Stability of the enzyme is not compromised by the purification procedures.
6. Low cost—no need to purify enzyme, substrates, cofactors, and so forth.
7. Permits the detection of the effect of various groups of substances via their action on certain cell loci or the overall metabolism of the cell.
8. Longer useful lifetime.

One of the main disadvantages of whole-cell systems is created by the high diffusional resistance of the cell and subcellular membranes, which increases the response time of the sensor. The rationale for combining intact cell systems with suitable transducing mechanisms for biosensing is the requirement for sensitive, stable, simple, and inexpensive biosensing devices to measure substances that require complex biological mechanisms in order to exert a measurable effect [5].

4.6.1. Microorganisms

Microorganisms such as bacteria and fungi have for many years been employed for analytical purposes, such as indicators of toxicity or for the measurement of specific substances. The deleterious effects of, for example, heavy-metal cations such as Hg^{2+} and Pb^{2+} on cell metabolism (inhibition of growth, cell viability, substrate uptake), or other substances on enzymatic activity, respiration, or bacterial luminescence require whole-cell systems to be interpreted and measured. Likewise, estimation of the concentration of certain substances in biological fluids can be affected by one or several of the large arsenal of enzymes and enzymatic pathways found in microorganisms; for instance, *Arthrobacter nicotiana* metabolize short-chain fatty acids and, therefore, can be used for the determination of these substances in for example, milk [65]. The rate of respiration, or the rate of production of certain metabolites as a result of an enzymatic reaction, can give an indication of the initial concentration of a substrate due to its requirement for cell growth. Specialized metabolic pathways can be induced in microbial cells by growing them on medium so that the required analyte must be utilized for growth. *Pseudomonas* grown on media containing nitriloacetic acid (NTA) as the only carbon source, activates an enzymatic pathway to metabolize it. These induced cells can then be used in conjunction with a NH_3-sensitive electrode for the determination of NTA. Whole microbial cells provide enzymes having a relatively high operational stability.

The use of microorganisms in biosensor systems requires their immobilization around the transducer surface [66]. Free cells entrapped between a semipermeable membrane and an electrode surface provide a method whereby cells are immobilized very gently. This is a necessity when the whole-cell system is to be exploited. If only one or a few intracellular enzymes are to be used less gentle immobilization systems can be used. Cell entrapment within membranes made from carbohydrate polymers such as carrageenan, cellulose, and alginate has been employed. Cells may also be immobilized biospecifically using naturally occurring substances, such as lectins and avidin, which bind to molecules on the surfaces of certain cell species.

Microbial sensors are potentially reuseable, due to the ability to replenish the cells. The biocatalytic activity of some microbial biosensors can be restored by placing the spent electrodes into nutrient growth medium. New cells then grow in situ. However, because of the buildup of dead cells, the sensor eventually must be discarded.

Transducers used to create biosensors based on the microbial reactions include ion-selective electrodes, p_{O_2}–electrodes, p_{CO_2}–electrodes, p_{NH_3}–electrodes, conductivity meters, thermistors, photometers, and piezoelectric membranes. Oxygen-sensitive electrodes have been commonly employed because of the large number of sensing systems based on the metabolic activity of the microorganism. Increases in respiratory activity are normally caused by assimilation of substrates (analytes) and are recorded as a decrease in O_2 tension. Although such systems give reliable results, they do not discriminate between different entities of the metabolic network. Tan et al. [67] developed a microbial sensor for the measurement of biochemical oxygen demand (BOD) in water samples, which consisted of a membrane containing the immobilized microorganisms, *Bacillus subtilis* and *Bacillus lichenformis,* to effect the biooxidation process and an electrochemical oxygen probe. The selectivity of the microorganism is unimportant, and high biooxidation activity is required for a wide range of organics. The ability of the mixed microbial culture to biodegrade most organics, resulting in the consumption of oxygen due to microbial respiration, was measured by the oxygen electrode. This biosensor could indicate the level of pollution from the BOD in a few minutes compared to 5 days using the traditional method.

4.6.2. Mammalian Systems

The use of whole mammalian tissue slices or cells as the sensing element in biosensor devices offers similar advantages to those provided by microbial systems. The use of tissue slices as catalytic layers is very attractive because of their high stability, high levels of activity, and low cost when compared with isolated enzymes. Such systems generally have longer lifetimes, offering enzymes available in their naturally optimized form. Tissue slices can be easily immobilized to transducer surfaces by simple mechanical fixation using a semipermeable membrane or a nylon net. They can be additionally crosslinked for mechanical stability. Tissue slices and purified enzymes can be coimmobilized, thereby increasing the diversity of such biosensors.

Xiuli et al. [68] immobilized a slice of rabbit thymus tissue on a monofilament nylon mesh with glutaraldehyde that was in contact with a NH_3-gas-sensing electrode. Thymus tissue contains a high concentration of adenosine deaminase, which catalyzes the release of NH_3 from adenosine. This sensor yielded an excellent electrode response for the quantitation of adenosine in biological samples.

Alternatively, mammalian cells cultured in vitro can be used as the biological sensing element. Nakamura et al. [69] developed a rapid detection system for allergic reaction using rat basophilic leukemia (RBL-1) cells as the biological element and cyclic voltammetry as the transducing mechanism. The cultured cells were immobilized to a membrane by filtration. When DNP was added to anti-DNP IgE-sensitized RBL-1 cells, it triggered them to secrete serotonin and histamine, which participate in a variety of acute allergic and inflammatory reactions. Serotonin present in the cell is electrochemically oxidized and measured by cyclic voltammetric techniques. The peak current increased linearly with increasing allergen (DNP) concentration within 20 min. Such a sensor would be impossible to develop without the use of whole living cells.

4.6.3. Plant-Tissue-Based Sensors

Plant tissues also are effective catalysts as a result of the enzymatic pathways they possess. Plant tissue offers the advantage of low cost relative to mammalian tissue slices. Probably the most famous plant-tissue-based sensor is the "banana" electrode. Banana pulp tissue was immobilized on the gas-permeable membrane of a Clark-type oxygen electrode [70]. Polyphenol oxidase, an enzyme found at a high concentration in banana tissue, converts dopamine to dopamine quinone and then to melanin utilizing oxygen. The oxygen utilized in the reaction was measured using the oxygen electrode. Botrè et al. [71] employed the cotyledons of legumes to determine the concentrations of diamines in biological samples. Diamine oxidase found in this tissue oxidizes diamines, and the H_2O_2 produced is determined by amperometry. This system is simpler than the alternative methods of diamine measurement, which include electrophoretic, chromatographic, and radioimmunologic methods. The same tissue-based sensor was also used in combination with a lysine decarboxylase membrane for the realization of a hybrid enzyme electrode for the determination of lysine.

Many plant-based tissue sensors have been developed utilizing tissue slices from the growing parts of plants, such as flowers and leaves [72], and the nutrient storage parts of plants [73].

4.6.4. Cell Receptors and Biosensors

Receptors are proteins located in the surface membranes of cells. Mammalian cells utilize receptors for the transmission of signals across the lipid bilayer membrane that separates intracellular and extracellular regions of the cell. In vivo, the binding of a ligand to its receptor initiates a particular physiological response in the receptor-bearing organism through activation or blockage of a series of biochemical reactions. Such events include hormone regulation, viral infection, and neural transmission. An example of such a receptor–ligand interaction is the binding of acetylcholine, a neurotransmitter, to its receptor in the sensitive ending of postsynaptic neurons, stimulating them to transmit an impulse. The binding of ligand to receptor results in a measurable response that is a representation of the physiological potency of the analyte.

Assays based on immunologic interactions are similar to receptor-based assays in that a binding reaction is involved. However, when both active and inactive forms of the analyte are present in a test sample, assays based on immunologic principles are defective in that physiologically inactive molecules may still bind to antibodies raised against them. In doing so, immunologic assays measure the total quantity of analyte present but give no indication of the biological activity. In receptor–ligand-based assays, intact biological transducers convert the extracellular binding reaction into an intracellular signal on the basis of either a transmembrane ion channel that opens or a membrane enzyme that undergoes activation or inhibition as a result of ligand binding. Only active forms of the analyte can activate such biological transducers. Receptor–ligand-based assays are applicable to a somewhat restricted but important range of metabolites such as steroid and protein hormones, neurotransmitters, certain drugs, and their active metabolites. However, tissue extracts are often less stable than antibodies and may contain receptors with other binding specificities.

One approach to the utilization of receptors in biosensors is in their isolated form in conjunction with a transducer. Such sensors are based on the specific recognition of the ligand for its receptor and not on bioactivity. The main problem with isolated receptors is stability, and many isolated receptors are also labile at ambient temperature. Some receptors such as cholinergic receptors require the presence of phospholipids in order to function. Generally, receptors must be reconstituted using phospholipid bilayers. The area of receptor immobilisation is discussed by Taylor [43].

Acetylcholine receptors were immobilized on the gate of an ion selective FET (field-effect transistor) by Gotoh et al. [74] and the potential measured with respect to another FET which did not have the receptor immobilised to it. The signal was due to the specific binding of acetylcholine to its receptor. Rogers et al. [75] determined the binding kinetics of the acetylcholine receptor for three fluorescently tagged neurotoxic peptides in an optical biosensor based on the evanescent wave technique.

An alternative approach is to use the receptors in their natural form, as intact tissue slices or in isolated whole cells. Such intact biological systems allow the ligand–receptor binding reaction to be transduced directly by the biological component and also helps to overcome the problems of receptor immobilization and stabilization. Intact biological transducers allow the conversion of chemical information into electrical impulses in a matter of milliseconds that can be displayed by means of, for example, an oscilloscope. Belli and Rechnitz [76] developed a biosensor using intact tissue. They used antennular structures from blue crabs that were sensitive to amino acids in solution and determined the concentration of the amino acids by studying the frequency of the potential spikes displayed on an oscilloscope. Other developments were made utilizing different chemosensing organs from different species sensitive to different amino acids, hormones, drugs, toxins, and neurotransmitters [77].

There is also a class of binding molecules that are not strictly biological receptors. Binding pairs such as biotin–avidin, protein A-IgG [78] and yeast mannan–concanavalin A [79], represent protein–ligand interactions. These binding proteins have limited but useful applications. They are more robust than biological receptors with regard to immobilization, and, therefore, adsorption and covalent binding is usually employed.

4.7. DNA/RNA-BASED SENSORS

Like the binding exhibited by antibodies, the interchain binding in nucleic acids is extremely specific. This fact can be exploited to use DNA and RNA as capturing agents in a biosensor system. Despite the inherent fragility of the DNA molecule, the increasing awareness of the importance of genetic factors in disease has given rise to a great deal of interest in the construction of biosensors that could detect regions of the genetic code.

4.7.1. Introduction to Structure and Function of DNA/RNA

The nucleic acids are responsible for storing and transmitting information in cells. They instruct the cells as to which proteins to synthesize, and in what amounts. Deoxyribonucleic acid (DNA) and ribonucleic acid (RNA) are the building blocks of genetics. DNA makes up the genes of all animal and bacterial cells and some viruses, while RNA is the genetic component of some other viruses.

DNA is a polymer, consisting of a chain of deoxyribonucleotide molecules joined together. A deoxyribonucleotide is composed of a sugar molecule (lacking an oxygen atom present in ribose—thus *deoxyribose*), one or more phosphate groups, and a nitrogenous base. Four different bases are found in DNA molecules: adenine (A), guanine (G), thymine (T), and cytosine (C). The deoxyribonucleotide units that contain these bases are called, respectively, deoxyadenylate, deoxyguanylate, deoxythymidylate, and deoxycytidylate.

Deoxyribose sugars joined together by phosphate groups form the "backbone" of DNA, which is constant throughout the entire molecule. Only the sequence of bases in the structure is open to variation—and it is this sequence that carries the genetic information. The sugars and phosphates are joined together by means of a phosphodiester bond, in which the 3'-hydroxyl group of one sugar is linked to the 5'-hydroxyl group of another. This method of linkage confers polarity on the DNA molecule, as one end of the chain will have an unlinked 5'-hydroxyl group and the other end, a free 3'-hydroxyl group. By common usage, the series of bases in a DNA chain is written in the direction of 5' to 3'.

The three-dimensional structure of DNA was described in 1954 by James Watson and Francis Crick. The molecule is in the form of a double right-handed helix, in which the two polynucleotide chains are wound around a common axis. The two chains are antiparallel—that is, they run in opposite directions, one 5' to 3' and the other 3' to 5'. The backbone of the molecule is on the outside of the double helix, with the bases on the inside. It is the bases that hold the structure together, as they bind with each other in a strictly defined way. Adenine always pairs with thymine, and guanine with cytosine. The bases are joined together by hydrogen bonds. An adenine–thymine pair has two such hydrogen bonds, while a guanine–cytosine pair has three. The helix has a diameter of 10 Å, and each turn contains 10 bases, which are spaced along the helix every 3.4 Å. This form of DNA is known as *B-DNA*, and is the type of DNA present in almost all cells under physiological conditions.

Rupturing the hydrogen bonds between the bases cause denaturation of the DNA molecule. One way to do this is by heating DNA solutions or adding acid or alkali to ionize the bases. Each DNA molecule has a distinctive "melting temperature" (T_m), defined as that temperature at which half of the double helix is unwound. The T_m of a particular DNA chain will depend on its base composition. Large numbers of GC pairings will reinforce the helical structure, giving the DNA molecule a T_m greater than DNA containing more AT pairings. This is due to the greater number of hydrogen bonds holding together a GC base pair. On reducing the temperature below the T_m, the separated strands of DNA will spontaneously recombine—in a process called *annealing*.

Because of the base-pairing restrictions, one strand of the double helix will always be the complement of the other strand. In other words, where a thymine is found on one strand, the other must have an adenine, and so on. This fact allows one of the strands to act as a template for the production of the other during DNA replication. In DNA replication, the mechanism is semiconservative—specifically, each daughter DNA molecule contains one strand from the parent molecule, and one that has been newly synthesized. Before replication can occur, the hydrogen bonds must first be broken, and the double helix unwound. A group of enzymes called the *DNA polymerases* then catalyzes the synthesis of a complementary strand to each parent strand. They do this by adding on deoxyribonucleotide units joined together by phosphodiester bonds, but only if they are complementary to the bases on the parent strand. Polymerases can also remove mismatched nucleotides from a newly synthesised chain, ensuring a high fidelity of replication.

In the cells of higher lifeforms, DNA is present in the form of chromosomes. The DNA in these chromosomes is tightly bound to a group of small proteins called *histones,* which account for up to 50% of the mass of the chromosome. Most of the DNA is wound around these proteins, forming units known as *nucleosomes.* Yeast cells contain 12–18 chromosomes, whereas human cells have two sets of 23 chromosomes. Most of the cells in the organism are *diploid*—they contain two sets of each chromosomes, although egg and sperm cells are *haploid,* containing only one copy of the genetic information.

RNA (ribonucleic acid) is, like DNA, a polymer consisting of nucleic acids, joined together by phosphodiester bonds. It differs from DNA, however, in that it contains ribose rather than deoxyribose as its sugar component. In addition, in place of thymine it contains the base uracil (U), which can pair with adenine but lacks a methyl group present on thymine. RNA is not capable of forming the double-helix characteristic of DNA, due to steric hindrance imposed by the ribose sugar units. It is usually single-stranded. A few RNA molecules, however, do contain some double-helical regions. In these situations, the RNA can fold back on itself, forming "hairpin loops," in which the base-pairing is often inaccurate. The central dogma of molecular biology can be stated as follows: "DNA makes RNA makes protein." Basically, the DNA molecule acts as the template for the production of a type of RNA called *messenger RNA* (mRNA). This process is known as *transcription.* One RNA molecule is produced for every gene expressed. The production of functional proteins from RNA is called *translation.* The instructions for protein synthesis are carried by nucleic acids in the form of a code. A sequence of three bases in the nucleic acid—known as a *codon*—specifies a particular amino acid. Codons in DNA are copied onto mRNA, which moves out of the nucleus of the cell. Another type of RNA, called *transfer RNA* (tRNA), reads the codons on the mRNA. At least one type of tRNA exists for each of the 20 amino acids. Structures in the cell called *ribosomes* join together the amino acids coded for to make proteins. The major component of ribosomes is another kind of RNA called *ribosomal RNA* (rRNA).

4.7.2. Practical Applications and Potential of DNA-Based Biosensors

The property of base complementarity makes possible the construction of DNA probes, which can be used in analytical techniques. If the sequence of bases composing a certain part of the DNA molecule is known, then the complementary sequence can be synthesized. This complementary sequence is called a *probe.* The probe can then be labeled. By unwinding the double helix (e.g., by addition of alkali), adding in the probe, and then annealing the strands, the labeled probe will

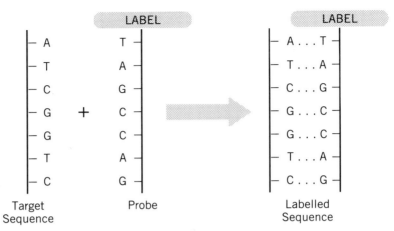

Figure 4.14. Diagrammatic representation of the action of nucleic acid probes. The target DNA double helix is unwound by heating or other means. The probe—a labeled complementary sequence of nucleic acids—is then added, and binds to the target sequence, forming a labeled, hybridized probe–DNA complex.

bind, or "hybridize," to its complementary sequence on the target molecule (Figure 4.14). Probes can be constructed by adding activated monomer units to a growing chain of DNA, which is immobilized on a solid support. In addition to this chemical method, molecular biological means are also available for the production of DNA probes. One approach is to remove mRNA for the gene of interest from cells in which it is present in abundance. By using an enzyme called *reverse transcriptase,* DNA that is complementary to this mRNA can be made in vitro and a specific probe produced. Another option is to discover the amino acid sequence of the protein coded for by the gene. The base sequence of the DNA can then be inferred, and a complementary probe produced. The problem with this method, however, is that the genetic code is *degenerate*—in other words, each amino acid is coded for by more than one codon.

One application of these probes is as analytical tools in the diagnosis of hereditary disease. Often the genetic mutations in the base sequence of DNA that lead to these diseases is known. The condition can then be diagnosed using the complementary sequence as a probe. Infectious diseases can also be detected using probes. All strains of the infectious agent are considered to contain a common DNA sequence region, so that only one probe is needed. In this respect, the use of DNA probes may have an advantage over methods employing antibodies, which could fail to distinguish between different strains of the same disease-causing organism. Another area in which DNA probes may have an advantage over immunoassay procedures is the detection of viral infection. Viruses consist solely of DNA or RNA contained in a protein coat. They inject their genetic material into the target cell, and "hijack" the cell's ribosomes, causing them to produce the viral proteins. Immunoassay procedures rely on detecting the antibody response to the infection, whereas testing for the presence of viral genes directly by using nucleic acid probes may facilitate earlier diagnosis.

Previously, DNA probes were radioactively labeled. In recent years, however, a range of nonisotopic labeling methods have become available, making probes suitable for use in biosensors. Several qualities are required of the label, in addition to the properties laid down for antibody labelling. Because of the sandwich assay design of a hybridization, and the high binding constant for the two DNA strands, the limit of detection of the label will generally be the limiting factor of the assay. Thus, the label chosen must be detectable at very low limits, and must not obstruct the binding of the probe and target DNA strands. Modification of the probe must also not lower its melting temperature, which would affect its hybridization properties. This means that the labeling should be restricted to sites on the DNA molecule not involved in hydrogen binding. Usually, however, hydrogen-binding amino groups are used, but only a small percentage of these are modified, so stability should not be affected. Enzymes have become popular choices for labeling, due to the range of methods that may be used for their detection. Chemiluminescent and fluorescent labels have also been widely used.

Labeling methods for DNA probes can be divided into chemical, synthetic, and enzymatic procedures. The range of functional groups available on a nucleic acid probe for chemical labeling is not wide. Synthetic labeling involves adding labels, or reactive groups suitable for labeling, onto the probe at the chemical stage. A variety of enzymatic methods have been described. Direct or indirect labeling methods may be employed. In the *direct* approach, the label is linked straight onto the nucleic acid by means of a covalent bond, or it binds noncovalently between the double-stranded probe/target sequence structure. *Indirect* labeling procedures, on the other hand, employ a hapten attached to the probe. For example, biotin may be used as the hapten, and bound to the probe. The biotin can then be detected by using a labeled binding protein such as avidin. Alternative, more complicated, indirect labeling strategies have been developed—including the introduction of compounds between hapten and labeled protein, and the use of binding proteins specific for double-stranded DNA.

Immobilization methods exist for a range of different biosensor applications using nucleic acid probes. Nucleic acids may be immobilized on a fluorocarbon surface for incorporation into biosensors. The fluorocarbon surface [e.g., polytetrafluoroethanol (PTFE)] is first treated with an activating composition consisting of a reactive polyfluoroalkyl sugar reagent. The activated surface is then treated with the DNA solution, in the presence of surfactant. Photoaffinity labels and photolithography can also be used to immobilize DNA/RNA probes. Another alternative is the coating of gold and silver surfaces with active monolayers of avidin or streptavidin. Biotinylated probes can then be linked to these sensor surfaces. A piezoelectric biosensor using a single-stranded DNA probe has been described [80]. The piezoelectric crystal was first coated with poly(butylmethacrylate) in ethyl acetate and dimethoxyphenylacetophenone in acetone. DNA was then covalently bound to the surface by irradiating with 365-nm ultraviolet light. The resonance frequency of the probe was then measured. On exposure to the target DNA sequence (denatured by heating to 100°C and then cooling to 65°C), hybridization occurs. The resonant frequency is again determined, and the degree of hybridization is calculated from the difference between the two resonances. This sensor format has been used to detect *Salmonella typhimurium* DNA in food.

4.8.   CASE STUDY: BIAcore™

An exciting development in the field of biosensors has been the advent of sensing technologies based on the optical phenomenon of surface plasmon resonance (SPR). Several commercial systems for the detection of biomolecular interactions using this method exist, of which the most widely used is BIAcore (real-time *biospecific interaction analysis*) from Pharmacia Biosensor AB.

The *principle* of BIAcore is as follows. If light is shone through a prism at an angle greater than the critical angle of reflection, then total internal reflection will occur—all the light will be reflected through the prism, with no refraction. If there is an interface between a medium of higher refractive index (e.g., the glass of the prism) and a medium of lower refractive index, then an electromagnetic field component of the light, termed the "evanescent wave," will penetrate a small distance into the medium of lower refractive index (Figure 4.15). The use of this evanescent wave to excite fluorophores in fiber-optic immunoassay systems has already been described in Section 4.3.7.2. If the interface between the two media is coated with a thin layer of metal film and the light shone on the prism is monochromatic and plane-polarized, then the delocalized surface electrons (*plasmons*) in the metal film will interact with the evanescent wave. At a specific angle of the incident light, the plasmons will become excited and absorb energy. A sharp dip in the intensity of the reflected light will be seen at this "SPR angle" [81]. This phenomenon is important because the SPR angle is dependant on the bulk refractive index of the medium on the metal-film side of the interface. The refractive index, in turn, is dependant on concentration. If the lower refractive index medium is buffer solution, then any changes in the concentration of the buffer will cause a change in the refractive index, and thus a change in the SPR angle.

BIAcore uses the combination of SPR analysis and continuous-flow technology to produce a sensor system by which interactions between molecules can be monitored in real time without labeling. Biomolecular interactions are studied by

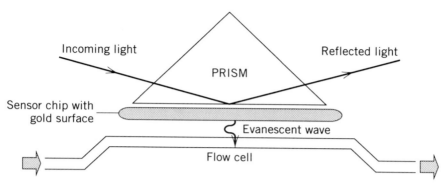

Figure 4.15. Surface plasmon resonance and BIAcore. The passage of light through the prism generates an evanescent wave. Delocalized surface electrons—or "plasmons"—in the gold film interact with
this evanescent wave, causing a dip in the intensity of reflected light at a specific angle. This surface
plasmon resonance (SPR) angle is measured by a photodiode array. BIAcore makes use of the fact that
this SPR angle is dependent on the concentration on the metal-film side of the interface. Biomolecules
are linked to the gold film and, by the use of continuous-flow technology, the interaction between the
immobilized molecule and sample passed over it can be monitored in real time.

immobilizing one of the components on the surface of a specialized sensor chip
within the biosensor, and allowing the other to flow over this surface in solution.
The system employs a gold-coated sensor chip as the metal-film component, in
conjunction with a prism and a high-intensity LED that emits a wedge-shaped
beam of light in the near-infrared region. The gold film on the sensor chip is derivatized with a long-chain hydroxyalkyl thiol monolayer, onto which is coated
100-nm-thick layer of carboxymethylated dextran. The carboxyl groups on this
dextran gel enable biomolecules to be immobilized using a range of different
chemistries. The dextran strands are mobile in three directions, and as such they
approximate more closely to a volume than an area. Immobilized biomolecules are
thus, in effect, still in solution.

The solution containing the second biomolecule is delivered to the sensor chip
surface by means of an autosampler and an integrated microfluidics cartridge
(IFC). The IFC consists of a series of precision-cast channels in a hard polymer
plate, which form sample loops and flow channels for delivering both buffer and
sample to the chip. Channels in the IFC are opened and closed by pneumatic mi-
crovalves, which are controlled by microcomputers. Two identical sets of loops
and channels are present on the IFC, and wash and injection cycles are alternated
between the two, eliminating the need for washing time in between samples. Four
parallel flow cells are formed on the sensor chip surface when the chip is "docked"
with the IFC. The wedge-shaped beam of light from the LED is shone simultaneously on all four flow cells and the reflected light is detected by a two-dimensional

diode array. The change in SPR angle is continuously monitored, and measured in arbitrary *resonance units* (RU's); 1000 RUs correspond to a 0.1° shift in SPR angle. For most proteins, such a signal is equivalent to a concentration at the chip surface of 1 ng/mm^2. The change in resonance angle is plotted against time and displayed as a sensorgram.

Several different approaches exist for immobilization of biomolecules onto the sensor chip surface. The most widely used method is derivitization with N-hydroxysuccinimide (NHS), mediated by N-ethyl-N'-(dimethylaminopropyl)carbodiimide (EDC). The dextran gel is first activated with EDC, and addition of NHS results in the formation of an NHS ester. This is a powerful leaving group, and will react readily with uncharged primary amino groups in biomolecules. An alternative approach is to immobilize via thiol–disulfide exchange. If no intrinsic thiol groups are present on the molecule, then reactive groups may be introduced by modifying amino or carboxyl groups. It is also possible to link molecules to the sensor chip by means of avidin–biotin binding.

The use of BIAcore has several clearcut advantages over more traditional techniques. The main advantage is the absence of labeling requirements, which often means that there is no need to purify the reactants. As the binding processes are observed in real time, the kinetic data (association and dissociation rates) are directly obtainable. A wide range of analytes can be measured, including drugs, antibodies, and nucleic acids. BIAcore can also be used for concentration measurement, and is effective over a large dynamic range of concentrations.

A sensor system has been described for the detection of HIV-1 using BIAcore [82]. A protein called gp160, which is a component of the protein envelope of the human immunodeficiency virus, is thought to be involved in the pathogenic process. Various peptide components of this protein were immobilized on the sensor chip by NHS esterification. Polyclonal human serum was obtained from HIV-positive volunteers. These serum samples were first heat-inactivated, and then allowed to interact with the peptide immobilised on the sensor chip. Once the binding had been measured, the peptide surface was regenerated by dissociating the bound antibody using dilute acid or base. Thus, it was possible to reuse the same peptide surface up to 100 times. The BIAcore signal was directly related to the concentrations of antigen and antibody within the biosensor.

BIAcore has also been used to study nucleic acid hybridisation [83]. The sensor chip was coated with avidin, which served as the reaction surface. A biotinylated nucleic acid probe sequence was then passed over the chip, and captured by means of avidin–biotin binding. Complementary DNA strands were then passed over the immobilized probe, and the hybridization was monitored on a sensorgram. Using this system, hybridization could be observed in less than 10 min at room temperature. The hybridization was also shown to be highly specific. It is hoped that the use of BIAcore for hybridization studies in the future will facilitate diagnosis of hereditary disease, DNA binding studies, and function analysis.

4.9. THE FUTURE

The exploitation of biospecific interactions has a key role to play in the development of highly sensitive biosensors with clinical, pharmacological, and biotechnological applications. Recently published material identifies this area as one of the fastest growing technologies with diverse applications in many scientific fields. Many of these developments rely on BIAcore technology, which offers several advantages over traditional techniques [84]. Recent developments include abzyme prodrug therapy [85], development of probes for viral and bacterial diagnosis [86–89], and the ability to analyze the molecular and kinetic parameters of antibody–antigen interactions [84,90–92]. New developments have also allowed the production of "tailormade" antibodies and the use of DNA probes as highly sensitive analytical reagents. The challenge for the future is to combine biomaterials with these developments and the associated developments in optical, electrochemical, and other methods of detection such that cheap, sensitive, and stable formats deliver reproducible results in a user-friendly manner in all potential applications. Ideally the sensor unit, with its associated biomolecules, should be in a disposable form that plugs into the transducer–detector unit. Judging by the advancement in biosensor technology in recent times, this goal should be achieved presently.

REFERENCES

1. Matthews, C. K., and van Holde, K. E., in *Biochemistry,* Benjamin/Cummins Publishing Co., Redwood City, CA, USA, 1990.

2. Zubay, G., in *Biochemistry,* 3rd ed., Wm. C Brown, Dubuque, IA, 1993.

3. Lehninger, A. L., in *Principles of Biochemistry,* Worth Publishing, New York, 1982.

4. Janata, J., *Principles of Chemical Sensors,* Plenum Press, New York, 1989.

5. Hall, E. A. H., in *Biosensors,* Hall, E. A. H., ed., Prentice-Hall, Englewood Cliffs, NJ, 1991, pp. 193–215.

6. Dixon, M., and Webb, E. C., *Adv. Prot. Chem.* **16,** 197 (1961).

6. Kohler, G., and Milstein, C., *Nature* **256,** 495–497 (1975).

7. Tracey, D. E., Liu, S. H., and Cebra, J. J., *Biochemistry* **15**(3), 624–629 (1976).

8. Ey, P. L., Prowse, S. J., and Jenkin, G. R., *Immunochemistry* **15**(7), 429–436 (1978).

9. Eveleigh, J. W., and Levy, D. E., *J. Solid Phase Biochem.* **2,** 44 (1977).

11. Koppel, G. A., *Bioconjugate Chem.* **1,** 13–23 (1990).

12. Carroll, K., Prosser, E., and O'Kennedy, R., *Technol. Ireland* 47–50 (1990).

12. Cooper, J. M., and McNeil, C. J., *Anal. Proc.* **27,** 95–96 (1990).

13. Killard, A., Deasy, B., O'Kennedy, R., and Smyth, M. R., *Trends Anal. Chem.* **14,** 257–266 (1995).

14. Kozbur, D., and Roder, J. C., *Immunol. Today* **4**, 72–79 (1983).

15. Barbas, C. F., III, Kang, A. S., Lerner, R. A., and Benkovic, S. J., *Proc. Natl. Acad. Sci. (USA)*, **88**, 7978–7982.

16. Old, R. W., and Primrose, S. B., *Principles of Gene Manipulation,* 4th ed., Blackwell, Oxford, 1992.

17. Innis, M. A., Gelfand, D. H., Sninsky, J. J., and White, T. J., eds., *PCR Protocols: A Guide to Methods and Applications,* Academic Press, London, 1992.

18. Graber, P., and Williams, G. A., *Biochim. Biophys. Acta* **10**, 193 (1953).

19. Kemeny, D. M., *A Practical Guide to ELISA,* Pergamon Press, London, 1991, p. 20.

20. O'Kennedy, R., *Clin. Chem. Enzymol. Commun.* **1**, 313–328 (1989).

21. Catt, K., and Tregear, G. W., *Science* **158**, 1570 (1967).

22. Campbell, A. M., in *Monoclonal Antibody Technology,* 4th ed., vol. 13 in *Laboratory Techniques in Biochemistry and Molecular Biology,* Burdon, R. H., and van Knippenberg, P. H., eds., Elsevier, Amsterdam, 1986, pp. 201–208.

23. Vaughan, A., and Milner, A., in *Antibodies,* Vol. 2, Catty, D., ed., IRL Press, Oxford, UK, pp. 201–222.

24. Johnson, G. D., in *Antibodies,* Vol. 2, Catty, D., ed., IRL Press, Oxford, UK, 1989, pp. 179–200.

25. Ligler, F. S., Shriver-Lake, L. C., and Ogert, R. A., in *Biosensors '92 Congress Proceedings,* Elsevier, Geneva, Switzerland, 1992.

26. Griffiths, D., and Hall, G., *TIBTECH* **11**, 122–130 (1993).

27. Schramm, W., Kuo, H., Hajizadeh, K., and Smith, R., in *Biosensors '92 Congress Proceedings,* Elsevier, 1992, Geneva, Switzerland.

28. Petersson, B. A., *Anal. Chim. Acta* **209**, 239 (1988).

29. Goding, J. W., *J. Immunol. Meth.* **20**, 241–253 (1978).

30. Akerstrom, B., and Bjorck, L., *J. Biol. Chem.* **261**, 10240–10247 (1986).

31. Eliasson, M., Olsson, A., Palmarantz, E., Wiberg, K., Inganas, M., Guss, M., Lindberg, M., and Uhlen, M., *J. Biol. Chem.* **263**, 4323 (1988).

32. Wilchek, M., and Bayer, E. A., *Anal. Biochem.* **171**, 1–32 (1988).

33. Hoshi, T., Anzai, J., and Osa, T., *Anal. Chem. Acta* **289**(3), 321–327 (1994).

34. Osa, T., Anzai, J., and Hoshi, T., *Abstr. Pap. Am. Chem. Soc.* 210 Meeting, Part 1, ANYLO14, 1995.

35. Malmqvist, M., *Nature* **361**(6405), 186–187 (1993).

36. Rishpon, J., *Biotechnol. Bridging-Res. Appl.* 95–107 (1991).

37. He, P. G., Takahashi, T., Anzai, J., Suzuki, Y., and Osa, T., *Pharmazie* **49**(8), 621–622 (1994).

38. Hoshi, T., Anzai, J., and Osa, T., *Anal. Chem.* **64**(4), 770–774 (1995).

39. Olson-Cosford, R. J., and Kuhr, W. G., *Anal. Chem.* **68**(13), 2164–2169 (1996).

40. Kanzaki, M., and Iwasawa, A., *Biomed. Res.* **16**(6), 381–386 (1995).

41. Basham, L. E., Pavlikar, V., Li, X. R., Hawwari, A., Kotloff, K. L., Edelman, R., and Fattom, A., *Vaccine* **14**(4), 439–445 (1996).

42. Guilbault, G. G., Kauffmann, J.-M., and Patriarche, G. J., in *Protein Immobilisation, Fundamentals and Applications,* Taylor, R. F., ed., Marcel Dekker, New York, 1991, pp. 209–262.
43. Taylor, R. F., in *Protein Immobilisation, Fundamentals and Applications,* Taylor, R. F., ed., Marcel Dekker, New York, 1991, pp. 286–289.
44. Robinson, G. A., Cole, V. M., Rattle, S. J., and Forrest, G. C., *Biosensors* **2,** 45–57 (1986).
45. Boiteaux, J. L., Desmet, G., and Thomas, D., *Clin. Chim. Acta* **88,** 329–336 (1978).
46. Aizawa, M., Morioka, A., and Suzuki, S., *Anal Chem.* **115,** 61–67 (1980).
47. Owaku, K., Goto, M., Ikariyama, Y., and Aizawa, M., *Sensors Actuators B,* **13–14,** 728–731 (1993).
48. Starodub, N. F., Arenkov, P. Y., Rachov, A. E., and Beregin, U. A., *Sensors Actuators B* **13–14,** 728–731 (1993).
49. Sandell, E. B., and Kolthoff, I. M., *J. Am. Chem. Soc.* **56,** 1426 (1934).
50. Greenwood, F. C., Hunter, W. M., and Glover, J. S., *Biochem. J.* **89,** 114–123 (1963).
51. Markwell, M. A. K., *Anal. Biochem.* **125,** 427–432 (1982).
52. Fraker, P. J., and Speck, J. C., *Biochem. Biophys. Res. Commun.* **80**(4), 849–857 (1978).
53. McFarlane, A. S., *Nature* **182,** 53 (1958).
54. Marchalonis, J. J., *Biochem. J.* **113,** 299 (1969).
55. Rosa, U., Scassellati, G. A., and Pennisi, F., *Biochim Biophys Acta* **86,** 519–526 (1964).
56. Bolton, A. E., and Hunter, W. M., *Biochem. J.* **133,** 529–539 (1973).
57. Hayes, C. E., and Goldstein, I. J., *Anal. Biochem.* **67,** 580 (1975).
58. Wood, F. J., Wu, M. M., and Gerhart, J. C., *Anal. Biochem.* **69,** 339–349 (1975).
59. Boiteaux, J. L., Desmet, G., Wilson, G., and Thomas, D., *Ann. NY Acad. Sci.* **613,** 390–395 (1990).
60. Raso, V., and Griffin, T., *Cancer Res.* **41,** 2073–2078 (1981).
61. Reading, C., in *Hybridomas and Cellular Immortality,* Tom, B. H., and Allison, J. P., eds., Plenum Press, New York, 1981, pp. 235–250.
62. Milstein, C., and Cuello, A. C., *Immunol. Today* **5,** 299–304 (1984).
63. Hollinger, P., Prospero, T., and Winter, G., *Proc. Natl. Acad Sci.* **90,** 6444–6448 (1993).
64. Blackburn, G. F., Talley, D. B., Booth, P. M., Durfor, C. N., Martin, M. T., Napper, A. D., and Rees, A. R., *Anal. Chem.* **60,** 2111–2116 (1990).
65. Ukeda, H., Wagnaer, G., Weis, G., Miller, M., Klostermeyer, H., and Schmid, R. D., *Z. Lebensm Unters. Forsch.* **195,** 1–2 (1992).
66. Karube, I., and Nakanishi, K., *Current Opinion Biotechnol.* **5,** 54–59 (1994).
67. Tan, T. C., Li, F., and Neoh, K. G., *Sensors Actuators B* **10,** 137–142 (1993).
68. Xiuli, G., Yugi, L., and Guanghua, Y., *Biosensors Bioelectron.* **7,** 21–26 (1992).
69. Nakamura, N., Kumazawa, S., Sode, K., and Matsunaga, T., *Sensors Actuators B* **13–14,** 312–314 (1993).
70. Sidewell, J. S., and Rechnitz, G. A., *Biotechnol. Lett.* **7,** 419 (1985).

71. Botrè, F., Botrè, C., Lorenti, G., Mazzei, F., Porcelli, F., and Scibona, G., *Sensors Actuators B* **15–16**, 135–140 (1993).

72. Uchiyama, S., and Rechnitz, G. A., *Anal. Lett.* **20**(3), 451–470 (1987).

73. Wang, J., and Lin, M. S., *Electroanalysis,* **1**(1), 43–48 (1989).

74. Gotoh, M., Tamiya, E., Momoi, M., Kagawa, Y., and Karube, I., *Anal. Lett.* **20**(6), 857–870 (1987).

75. Rogers, K. R., Eldefrawi, M. E., Menking, D. E., Thompson, R. G., and Valdes, J. J., *Biosensors Bioelectron.* **6**, 507–516 (1991).

76. Belli, S. L., and Rechnitz, G. A., *Anal. Lett.* **19**, 403–416 (1986).

77. Buch, R. M., Barker, T. Q., and Rechnitz, G. A., *Anal. Chim. Acta* **243**, 157–166 (1991).

77. Muramatsu, H., Dicks, J. M., Tamiya, E., and Karube, I., *Anal. Chem.* **59**(23), 2760–2763 (1987).

79. Janata, J., *J. Am. Chem. Soc.* **97**(10), 2914–2916 (1975).

80. Fawcett, N. C., *World Intellectual Property Organisation,* WO 87/02066, 1987.

81. Raether, H., *Surface Plasmons on Smooth and Rough Surfaces and on Gratings,* Springer, Berlin, 1988.

82. Vancott, T. C., Loomis, L. D., Rredfield, R. R., and Birx, D. L., *J. Immunol. Meth.* **146**, 163–176 (1992).

83. Wood, S. J., *Microchem. J.* **47**, 330–337 (1993).

84. Laricchie-Robbio, L., Liedberg, B., Platou-Vikinge, T., Rovero, P., Beffy, P., and Revoltella, R. P., *Hybridoma* **15**, 343–350 (1996).

85. Wentworth, P., Datta, A., Blakey, D., Boyle, T., Partridge, L. J., and Blackburn, G. M., *Proc. Natl. Acad. Sci.* USA **93**(2), 799–803 (1996).

86. MacKenzie, C. R., Hirama, T., Lee, K. K., Altman, E., and Young, N. M., *J. Biol. Chem.* **272**, 5533–5538 (1997).

87. Richalet-Sécordel, P. M., Poisson, F., and Van Regenmortel, M. H. V., *Clin. Diagn. Virol.* **5**, 111–119 (1996).

88. Myszka, D. G., Arulanantham, P. R., Sana, T., Wu, Z., Morton, T. A., and Ciardelli, T., *Protein Sci.* **5**, 2468–2478 (1996).

89. Scalia, G., Halonen, P. E., Condorelli, F., Mattila, M. L., and Hierholzer, J. C., *Clin. Diagn. Virol.* **3**(4), 351–359 (1995).

90. Diaw, L., Magnac, C., Pritsch, O., Buckle, M., Alzari, P. M., and Dighiero, G., *J. Immunol.* **158**, 968–976 (1988).

91. Myszka, D. G., *Current Opinion Biotechnol.* **8**, 50–57 (1997).

92. England, P., Bregergere, F., and Bedouelle, H., *Biochemistry* **36**, 164–172 (1997).

CHAPTER

5

OPTICAL CHEMICAL SENSORS

BRIAN D. MacCRAITH

BEST (Biomedical and Environmental Sensor Technology) Centre, School of Physical Sciences, Dublin City University, Dublin, Ireland

5.1. INTRODUCTION

Optical techniques for chemical analysis are well established. Sensors based on these techniques are now attracting considerable attention because of their importance in applications such as environmental monitoring, biomedical sensing, and industrial process control. In many instances these sensors exploit specific advantages of optical fiber technology. Fiber-optic chemical sensors (FOCS) can benefit from, for example, the geometric versatility, low attenuation, and electrical noise immunity of optical fibers. In this chapter, the emphasis throughout is on FOCS, although many of the methods described can be transferred, sometimes with considerable advantage, to planar wave guide configurations. Alternatively, they can operate successfully without the use of optical fibers at all. This latter operation is often neglected in the general promotion of optical chemical sensors.

A comprehensive study of the broad range of issues underlying FOCS is presented in the excellent two-volume text edited by Wolfbeis [1]. More recent updates on developments in the field are to be found in the proceedings of Europtrode I (1992), Europtrode II (1994), and Europtrode III (1996), all of which were published in special editions of *Sensors and Actuators B* [2]. In addition, the proceedings of the annual SPIE meeting on Chemical, Biochemical and Environmental Fiber Sensors provide a useful overview of research activity in the area [3]. Other important reviews are to be found in References 4 and 5.

In this chapter, an overview of FOCS is provided, and instrumentation, sensor design, and some specific applications are discussed. The objective is not to provide a complete literature review (this may be found in Refs. 1, 4, and 5) but rather to inform the reader of the general features of the technology. The general principles of operation of FOCS are first presented together with the conventional classification of the various configurations employed. The associated optoelectronic components such as sources, detectors, and optical fibers are also introduced. The principal sections of this chapter, however, deal with the design and operational issues for a range of FOCS configurations, and useful guidelines are provided in particular for those initiating research in this area. Those designs and applications with the greatest potential for eventual commercialization are highlighted. A number of issues that in the past have hindered the commercialization of FOCS and must be overcome if this research area is to be successful are identified. Overall, this chapter aims to provide the reader with a broad, balanced overview of the current state of FOCS technology, together with enough specific information to facilitate practical implementation of the ideas presented.

5.2.   PRINCIPLES AND ADVANTAGES OF FIBER-OPTIC CHEMICAL SENSORS

Chemical sensing based on optical fibers has many attractive features:

1. The technology has access to a multiplicity of optical techniques already developed for routine chemical analysis.
2. The low attenuation of optical fibers enables remote in situ monitoring of species in difficult or hazardous locations, such as groundwater monitoring or monitoring of process streams in nuclear fuel reprocessing plants.
3. These sensors can exploit the high-quality components (fibers, sources, detectors, connectors, etc.) developed for the more mature fiber-optic telecommunications technology.
4. The geometric flexibility of optical fibers and the feasibility of miniaturization may both be exploited in FOCS systems.

Fiber-optic chemical sensors are classified conveniently in two categories (Figure 5.1):

1. *Direct Spectroscopic Sensors.* In this case, the fiber functions solely as a simple lightguide that separates the sensing location from the monitoring instrumentation (source, detector, spectral filtering, etc.). The fiber facilitates direct spectral analysis (e.g., fluorescence, absorption, Raman scattering) of a sample at a distance. Naturally fluorescent groundwater contaminants have been monitored in this manner at sub-ppm levels over distances of hundreds of meters [6].

2. *Reagent-Mediated Sensors* (*Optrodes*). Here the optical fiber is combined with chemical reagents. A reagent is chosen to react sensitively and specifically to the analyte and the resultant change in its optical properties (e.g., fluorescence or absorption) is a direct measure of the analyte concentration. A number of configurations may be used. For example, at the far end of the fiber specific reagents are contained in a miniature reservoir attached to the fiber tip and are separated from the sample by means of an appropriate membrane. Alternatively and more usual-

(a) Direct Spectroscopic Sensors

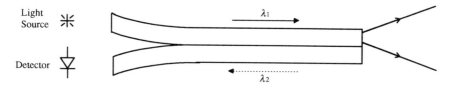

(b) Reagent-Mediated Sensors (Optrodes)

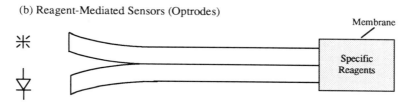

Figure 5.1. Classification of fiber-optic chemical sensing: (*a*) direct spectroscopic sensors; (*b*) reagent-mediated sensors (optrodes).

ly, suitable reagents may be immobilized directly in a support matrix on the fiber tip or along the core of a declad optical fiber. The term *optrode* is often used to describe reagent-mediated sensors and is derived from the combination "optical-electrode."

In addition, the terms *extrinsic* and *intrinsic* are sometimes applied to FOCS. Extrinsic fiber sensors are those in which the function of the optical fibers is to convey the light to and from the sensing location, which may or may not employ chemical reagents. The optical fibers play no other part in the sensing mechanism. In contrast, the optical fiber plays an active role in intrinsic sensors, insofar as the light transmitted is modulated by chemically induced interactions in the fiber core or, more usually, in the fiber cladding. The most common examples of intrinsic FOCS are those in which the fiber cladding is chemically sensitive and is interrogated via the evanescent field the guided light. Evanescent wave sensors are described in detail in Section 5.5.

5.3. INSTRUMENTATION

The generic FOCS consists principally of a light source, an optical detector, optical fiber(s) and, in the case of optrodes, immobilized reagents. Additional elements such as lenses, filters, and other spectral filtering devices are usually required to complete the system. This section concentrates purely on the optoelectronic elements of the FOCS system. Background details and suppliers of the relevant components are to be found in optoelectronics trade journals such as *Laser Focus World* and *Photonics Spectra*.

5.3.1. Optical Fibers

Light is guided in an optical fiber by total internal reflection at the core–cladding interface as shown in Figure 5.2. This mechanism requires the cladding refractive index n_2 to be lower than that of the core, n_1. The critical angle $\Theta_c = \sin^{-1}(n_2/n_1)$ at the core–cladding interface corresponds to the maximum external launch angle Θ_{max} that is compatible with light guidance in the fiber. The core diameter ($2a$) of optical fibers ranges from a few micrometers ($\mu m = 10^{-6}m$) in the case of single-mode fibers to over 1 mm in highly multimode fibers. The term *mode* refers to an allowed electromagnetic field distribution that propagates along the fiber; this description emerges from a rigorous electromagnetic field analysis of light propagation in optical fibers. A ray-optics approach to lightguiding, however, is valid for the large core diameters ($>100\,\mu m$) optical fibers that are most often used in FOCS.

   An important parameter in the design of FOCS is the numerical aperture (NA) of the fiber used. The NA is crucial, for example, in the calculation of the light-collection efficiency of a FOCS system and is given by

$$NA = n_0 \sin \Theta_{max} \qquad (5.1)$$

where Θ_{max} is the maximum launch angle defined earlier and n_0 is the refractive index of the medium from which light is entering the fiber. This medium is usually air, in which case $n_0 = 1$.

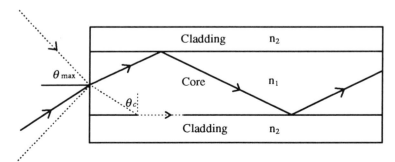

Figure 5.2. Lightguiding conditions in an optical fiber.

In the context of determining the number of modes that may exist in an optical fiber and also the extent of evanescent wave interactions at a particular wavelength λ, the fiber V number is an important parameter given by

$$V = \frac{2\pi}{\lambda} a\mathrm{NA} \qquad (5.2)$$

When $V < 2.405$, only one mode can propagate in the fiber and it is therefore termed *single-mode*. In the context of FOCS, the most important property of optical fibers, apart from those already mentioned, is the useful spectral transmission window of the fiber which is determined by the fiber attenuation. The fiber attenuation describes the signal loss per unit length along the fiber and is measured in decibels per kilometer or meter. (*Note*: 3 dB corresponds to a 50% power loss.) All-silica optical fibers, developed for telecommunications purposes, have the lowest attenuation (as low as 0.2 dB/km at $\lambda = 1550$ nm), and may be used for wavelengths between 350 and 1800 nm. For a variety of reasons most FOCS systems exploit nontelecom fibers. For example, polymer-clad silica (PCS) fibers are available with large core diameters (e.g., $600\,\mu$m) and are easily declad by chemical etchants for evanescent wave sensors. All plastic fibers are also available in both single-fiber and fiber-bundle format. The quality of these fibers has improved in recent years, and they are particularly suited to transmission of visible radiation over short distances. The benefits of the fiber-bundle format include the large area for efficient light launch and the ease with which a bifurcated configuration can be formed. This configuration enables the user to link the sample with both the light source and detector using a single-fiber bundle.

The spectral bandwidth of silica-based fibers renders them suitable only for visible and near-IR applications. In the case of sensors based on direct spectroscopic absorption, the midinfrared region ($\lambda > 2\,\mu$m) offers the advantages of much stronger absorption and much fewer overlapping bands than the near-IR region. For this reason there has been considerable interest in the development of fibers for the mid-IR region. Table 5.1 lists some of the fibers that have been produced together with their useful spectral window.

These fibers have been used for applications such as sensing of flammable gas-

Table 5.1. Properties of Some Mid-IR Optical Fibers

Material	Spectral Bandwidth (μm)	Attenuation[a]
Zirconium fluoride	0.5–5	0.002 dBm^{-1} at 2.55 μm
Chalcogenide (e.g., $As_x S_{1-x}$)	1–6	0.3 dBm^{-1} at 2.5 μm
Silver halide (e.g., $AgCl_x Br_{1-x}$)	3–15	0.5 dBm^{-1} at 10.6 μm

[a] dBm^{-1} = decibels per meter.

es (e.g., based on methane absorption at $\lambda = 3.4\,\mu m$), CO_2 monitoring at $\lambda = 4.3$ μm, and detection of chlorinated hydrocarbons in water at wavelengths in the vicinity of $10\,\mu m$.

5.3.2. Optical Sources

The selection of the appropriate light source is a crucial factor in the design of an optical chemical sensor system. The issues that must be considered include

1. Spectral output
2. Intensity
3. Stability
4. Ease of modulation
5. Predicted lifetime
6. Cost and power consumption
7. Size (and ease of coupling into optical fibers)

It is often the case, however, that the spectral requirements of a particular application restrict the selection of light source to a single option or at best a small number of choices. For example, in the ultraviolet region the only broadband sources available are deuterium and xenon lamps. In general terms, the light sources of relevance to FOCS may be classified under the following headings:

1. Incandescent lamps
2. Discharge lamps
3. Lasers (nonsemiconductor)
4. Semiconductor sources (LEDs and laser diodes)

1. *Incandescent Lamps.* Incandescent filament bulbs (such as tungsten halogen lamps) emit over a broad spectral range (visible, IR), are relatively inexpensive and are available in compact sizes. They are not, however, suited to modulation; they generate heat and have much shorter operational lifetimes than LEDs, for example. In addition, issues such as filament movement can give rise to alignment instability and consequent measurement inaccuracy.

2. *Discharge Lamps.* The UV range of 200–400 nm is important for both absorption measurements and excitation of fluorescence. Deuterium lamps are best suited to the 200–300-nm range but are generally bulky, are more expensive, and have higher power consumption than do incandescent sources. In addition, power supply requirements are more demanding and consequently more expensive. Xenon flashlamps provide a broadband output in the 200–1000-nm range but re-

quire power source regulation and suffer from pulse-to-pulse variation in output intensity.

3. *Lasers.* Disregarding semiconductor sources under this heading, a very limited selection of lasers is available at specific wavelengths in the UV–vis region of the spectrum, for example, Ar ion at 488 nm and 514 nm, HeNe at 633 nm, and HeCd at 325 nm and 442 nm. Lasers offer the advantages of monochromatic output (obviating the need for a spectral selection filter), high intensity, and directionality (facilitating launch into optical fibers, for example). These features, however, are often outweighed by cost and size considerations, with only the HeNe visible laser currently available at a cost less than $1500. Furthermore, these sources are fragile and some may drift out of alignment if subject to mechanical disruption.

4. *Semiconductor Sources.* Semiconductor solid-state sources that include laser diodes and LEDs are the most attractive option for FOCS because of their low power consumption, high stability, long lifetime, robustness, and compact size. In addition, such sources are usually inexpensive, especially at visible and near-IR wavelengths. Laser diodes, in particular, provide intense collimated beams and are easily modulated. However, unlike LEDs, these sources are not yet available at wavelengths below 630 nm, and this precludes their use in many interesting fluorescence-based sensing systems, for example. Wavelength regions occupied by routinely available laser diodes are 630–670 nm, 750–830 nm, and the "telecom windows" at 1300 and 1550 nm. Some laser diodes fabricated specially for near-IR spectroscopic chemical sensing applications are available commercially [7] but are expensive ($1000s). In the context of remote multipoint or distributed sensing there is growing research interest in the concept of combining near-IR dyes (absorbing and fluorescent) with low-cost laser diodes available in this region, especially in the vicinity of the first telecom window at 850 nm. This approach offers the advantages of low fiber attenuation and low background intrinsic fluorescence.

At the time of writing there is a very significant R&D effort under way in Japan, the United States, and Europe with the objective of developing room-temperature blue laser diode sources, based on GaN or ZnSe materials. The principal motivation for this work is the increased information storage capacity at blue wavelengths (e.g., 450 nm) for CDs and CD-ROMs. The routine availability of such sources would give a major stimulus to fluorescence-based chemical sensors and biosensors, facilitating the development of both compact and remote systems. Frequency-doubled near-IR sources that emit blue laser light are already available but are currently too expensive for most applications. Other laser sources that have been used in sensor applications include cryogenically cooled lead-salt diode lasers in the mid-IR region and rare-earth-doped fiber lasers. A thulium-doped fiber laser emitting at 2.3 μm has been used for hydrocarbon gas sensing [8].

Light-emitting-diodes (LEDs) are inexpensive semiconductor sources with relatively narrow emission bandwidths ($\sim$50 nm) as compared with incandescent sources. The range of LEDs now available spans the whole of the visible and near-IR spectra and many of these provide output powers in the milliwatt region. A recent significant development has been the production of high-intensity blue LEDs based on GaN materials. These have already been exploited in many fluorescence-based sensor systems [9].

In addition to the advantages of semiconductor sources already given, LEDs are particularly attractive sources for optical sensing because of their very low cost, ease of modulation, and ease of coupling to multimode optical fibers. Unlike lasers, they are not sensitive to backreflections and have low coherence. Both of these can cause serious problems in diode-laser-based sensing systems with the former being a potential cause of laser destruction and the latter giving rise the phenomenon of modal noise. Care must be taken, however, to correct for the temperature sensitivity of the LED emission, which typically undergoes a peak wavelength shift of 0.3 nm/K. In a typical application, a visible/NIR LED, modulated at an appropriate frequency, is used for absorption or fluorescence excitation. The transmitted/emitted signal is then detected by a silicon photodiode connected to a lock-in amplifier circuit. The wide range of available LEDs also facilitates the selection of an additional source at an adjacent nonabsorbed wavelength for the purpose of intensity referencing (i.e., compensating for non-measurand-related signal intensity changes in the measurement system).

An important development in the area of direct spectroscopic sensors based on infrared absorption in the 2–5 μm region is the recent availability of LEDs fabricated for this purpose. Although these first generation devices are more expensive than their vis/NIR counterparts, and require more complex drive electronics to achieve useful output powers, they provide a fiber-compatible source in a spectral region that hitherto relied on filtered incandescent or blackbody sources. Hydrocarbon gas sensing [10] has been demonstrated with these sources, which are available mainly from Russian suppliers [11].

5.3.3. Detectors

Where suitable, semiconductor photodiodes are the detectors of choice in most sensor applications. These are p–n junction devices that are generally operated in reverse-biased photoconductive mode in order to provide a linear response to signal intensity. Silicon photodiodes are low-cost detectors with a spectral response curve that spans the visible region but falls off sharply above 1000 nm. Avalanche photodiodes (APDs) may be used in applications where high gains are required. Photomultiplier tubes (PMTs) are significantly more expensive than photodiodes but offer much greater sensitivity and are required when dealing with very low light levels. In extreme cases single-photon-counting versions of PMTs may be used. PMTs are now available in compact units (matchbox size) with accompanying power supply modules that provide the high voltage for PMT operation while requiring only a low-voltage DC supply. Some suppliers also provide even smaller units in cans that are suitable for mounting on printed circuit boards.

In the IR region a much greater range of detectors is available although the selection is usually determined by the spectral range of interest. Rugged, low-cost photothermal detectors such as pyroelectric detectors find some limited use, but in most instances photoconductive detectors are used. The spectral operating range is determined by the semiconductor materials used in the fabrication of these devices; for example, PbS detectors are used in the 2-μm region, PbSe bewteen 3 and 5 μm, InSb between 5 and 7 μm, and HdCdTe between 5 and 14 μm. All of these devices operate best when cooled in order to reduce thermal noise. Usually, PbS and PbSe detectors are mounted on Peltier units that provide thermoelectric cooling typically down to a temperature of $-30°$C, although this depends on the number of stages used. Both the InSb and HgCdTe detectors are generally attached to an 8-h liquid nitrogen Dewar to provide cooling to 77 K. Such considerations clearly present problems where long-term unattended use of sensors is required. Consequently there is growing interest in the use of alternative methods of cooling such as Stirling microcoolers.

5.3.4. Spectral Selection and Other Components

Apart from fibers, sources and detectors, the other major instrumental consideration is the means of spectral selection where required. This may be necessary for both the input and output optical signals. In the simplest cases the use of appropriate filters is all that is required. Filter technology has improved significantly in recent years, and many sensor systems now employ a single dichroic filter for separation of excitation and fluorescent light or holographic edge filters for removal of the pump beam in Raman spectroscopic systems. A wide range of doped glass filters and interference filters is available for broadband applications. For narrowband requirements only multilayer interference filters are suitable, but care must be taken to avoid the problem of the thermal and orientational sensitivity of the filter passband.

In applications where wavelength tunability is necessary or where spectral information over a broad wavelength range is required, two options are available. First, compact scanning monochromator systems may be used although these are slow devices and may not lend themselves to robust treatment. Second, fiber-compatible compact CCD-array spectrometers are now available from a number of suppliers [12] and, although the typical spectral resolution is no better than a few nm, they provide spectra in real time and enable rapid multispectral processing of data. Moreover, their small size together with the absence of moving parts make them particularly suited to applications where portability is an advantage.

The other optical components that are required regularly in FOCS systems are lenses, mirrors, and optical connectors. These are all available routinely from optical component suppliers. It is worth noting that in laboratory development systems microscope objective lenses are generally used for launching light into optical fibers because of their specified numerical aperture, short working distance, and consequent small spot size. Fiber-optic couplers, which divide light from an incoming fiber in a specified ratio between two or more outgoing fibers, are often more convenient and more rugged than beamsplitters. Although couplers are available routinely for telecom-grade optical fibers, only a small number of specialist suppliers provide such components for large-core PCS fibers. It is also important to note that the splitting ratio in couplers is usually temperature-sensitive and wavelength-dependent.

5.4. SPECTROSCOPIC PRINCIPLES

Most optical chemical sensors are based on a spectroscopic technique such as measurement of absorption or fluorescence, whereby the detected signal is used to deduce the concentration of the target analyte. It is important, therefore, to review the basic principles and limitations of the various spectroscopies used. The principle spectroscopies used in optical chemical sensors are illustrated schematically in Figure 5.3, which shows an incident beam of intensity I_0 yielding transmitted (I_T), reflected (I_R), or fluorescence (I_F) signals, after interaction either directly with the analyte or with an immobilized indicator system. Although a range of geometrical configurations can be used, with or without optical fibers, the basic principles of interaction remain the same.

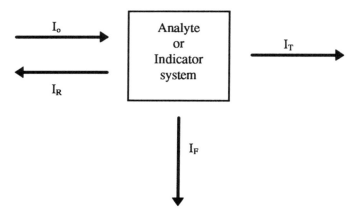

Figure 5.3. Major spectroscopic principles used in optical chemical sensors.

5.4.1. Absorption

Absorption is characterized by the Beer–Lambert law:

$$I_T = I_0 \, 10^{-\epsilon L C} \tag{5.3}$$

where ϵ is the molar absorptivity (L mol^{-1} cm^{-1}) of the absorbing species, L is the absorption path length (cm), and C is the concentration (mol/L) of the absorbing species.

Some spectroscopists prefer to use the equivalent expression

$$I_T = I_0 \, \exp(-\alpha L) \tag{5.4}$$

where α is the absorption coefficient (uin cm^{-1}) of the absorbing material. A modified version of Eq. (5.4) is particularly useful in describing evanescent wave absorption sensors (see Section 5.5).

The absorbance of the sample or indicator system is given by

$$A = \log_{10}\left(\frac{I_O}{I_T}\right) = \epsilon L C \tag{5.5}$$

The most important practical consequence of this equation is that when I_T and I_0 are known, the calculated absorbance A is directly proportional to the concentration C, thereby yielding a linear calibration.

If Eq. (5.4) is used, it is easy to show that

$$A = 0.434 \, \alpha \, L \tag{5.6}$$

It is important to point out, however, that there are practical limitations to the applicability of the Beer–Lambert (BL) law, or, in particular, the linear dependence of absorbance on concentration:

1. The bandwidth $\Delta\lambda$ of the incident beam should be very narrow, ideally approximating monochromatic radiation. Deviations from perfect BL behavior increase as $\Delta\lambda$ increases, but are particularly severe when $\Delta\lambda$ is greater than the spectral width of the absorption band of the absorbing species. Such deviations result in a nonlinear calibration for the sensor, which, although acceptable in some instances, results in a sensitivity that falls off with concentration and a consequent reduction in the useful measurement range.

2. Deviations from perfect BL behavior are also observed in highly absorbing or highly scattering media. Both of these effects yield a very limited linear range for the absorbance–concentration relationship. Furthermore, high

concentrations of the absorbing species can also result in measurement problems due to reactions or complexation taking place. In such circumstances the optical characteristics of the absorbing material will differ significantly from those observed at low concentrations.

5.4.2. Reflectance

Diffuse reflectance from an optically rough layer of absorbing species may also be used to deduce the concentration C of that species. As with absorbance, this may yield the concentration of the analyte directly or indirectly. The reflectance R $(= I_R/I_0)$ is related nonlinearly to C via the Kubelka–Munk function F_{KM}, which is given by

$$F_{KM} = \frac{(1-R)^2}{2R} = \frac{\epsilon C}{S}$$

where S is a scattering coefficient. Although diffuse reflectance spectroscopy is a well-established analytical technique with particular application in near-IR analysis of powders and other opaque materials, the use of this technique in FOCS has been very limited.

5.4.3. Fluorescence

Fluorescence is the radiative deexcitation of a molecule flowing absorption of a photon. In general, the emitted photon is of lower energy than the absorbed photon. Consequently, the fluorescence emission peak of a species is at a longer wavelength than the absorption peak. The wavelength separation between these peaks is called the *Stokes shift*.

In the context of the parameters already defined in this section, it is easy to show that

$$\log_e \left(\frac{I_T}{I_0} \right) = -2.303 \epsilon L C \tag{5.7}$$

Therefore,

$$I_T = I_0 \exp(-2.303\ \epsilon L C)$$

$$= I_0 [1 - (2.303)\ \epsilon L C]$$

when $\epsilon L C$ is small ($\ll 1$). One can then show that

$$I_0 - I_T = I_0 \, (\epsilon LC) \, (2.303) \tag{5.8}$$

In the absence of significant scattering, this quantity $(I_0 - I_T)$ is simply the intensity of light absorbed by the sample, and one expects that the fluorescence intensity I_F is linearly proportional to this quantity. Therefore,

$$I_F \propto I_0 \, \epsilon LC \tag{5.9}$$

This linear dependence of the fluorescence emission intensity on the concentration of the absorbing species, in particular, is the underlying principle of fluorometric sensing. Clearly the derivation of Equation (5.8) requires that the product ϵLC be small ($\ll 1$). At high concentrations of the emitting species, the linear relationship expressed in Eq. (5.8) breaks down as shown in Figure 5.4. In addition, the problematic phenomena of self-quenching and self-absorption (inner filter effect) begin to occur as the concentration C is increased. Although some optical sensors are based on the intrinsic fluorescence of the analyte, most are based on the reagent-mediated approach. In these latter applications, it is clearly important, therefore, to use a sufficiently low reagent concentration in order to eliminate the possibility of encountering the problems outlined above. Care should be also taken to avoid *photobleaching,* which is the irreversible photodegradation of the fluorescent species, due to an excessively high excitation intensity I_0.

A particularly important class of fluorometric optical sensors is that based on

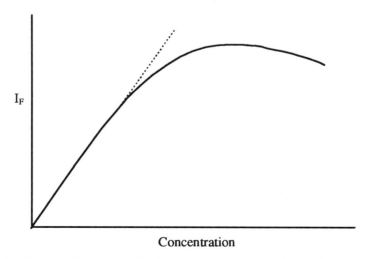

Concentration

Figure 5.4. Variation of fluorescence signal with fluorophore concentration showing onset of nonlinearity at higher concentrations.

fluorescence quenching, whereby excited reagents are deexcited nonradiatively by collision with specific analyte molecules. This process is described by the Stern–Volmer equation:

$$\frac{I_F^{(0)}}{I_F} = 1 + K_{SV}[Q] \tag{5.10}$$

where $I^{(0)}_F$ and I_F are respectively the fluorescence signals in the absence and presence of quencher, $[Q]$ is the quencher concentration, and K_{SV} is the Stern–Volmer quenching constant. This constant is given by

$$K_{SV} = k\tau_0 \tag{5.11}$$

where k is the diffusion-dependent bimolecular quenching coefficient and τ_0 is the fluorescence decay time of the species in the absence of the quencher. Two important practical consequences of Eq. (5.10) are as follows:

1. The sensitivity of the quenching process is enhanced by employing reagents with long fluorescence decay times.
2. The sensitivity of the process may be tailored by controlling the quencher diffusion rate via the microstructural properties of the immobilization matrix.

5.4.4.   Absorption versus Fluorescence

In general, when the two options of absorption and fluorescence are available for a particular sensing application, it is better to choose fluorescence for the following reasons:

1. Fluorescence yields a signal that usually increases with analyte concentration and is measured against a zero background light level (ideally). In contrast, absorption is based on measurement of a signal that decreases with increasing analyte concentration and is measured with respect to a high background. Intrinsically, therefore, the fluorescence approach will yield higher signal-to-noise ratios than absorption does.

2. In the case of fluorescence, one measures light that has been emitted by analyte or reagent molecules. Consequently, it contains much useful information about these molecules. For example, one can measure fluorescence intensity, decay time, and polarization anisotropy and can also detect the phenomenon of energy transfer. Absorption, on the other hand, is based on the measurement of resid-

ual transmitted light, which contains little information about the absorbing molecule.

5.4.5. Other Spectroscopies

Although absorption and fluorescence are the dominant spectroscopic interactions on which optical chemical sensors are based, other spectroscopies are very useful in particular applications. Raman spectroscopy, for example, may be used to yield "fingerprint" vibrational spectra of species. It can be used for both quantification and compositional identification. Because it is based on a very weak scattering phenomenon, laser sources and sensitive detectors are generally required. In recent years, however, compact systems based on diode lasers, holographic notch filters (for laser signal rejection), and detector arrays are emerging. These may well prove to be useful in industrial process monitoring and control applications. It is possible in some circumstances to enhance the weak Raman effect by orders of magnitude when the sample is located on metal island films. This so-called surface-enhanced Raman spectroscopy (SERS) is the subject of increasing research interest. However, it is based on principles not yet fully understood, and has yet to yield reliable, reproducible sensors.

Finally, some chemical/biochemical reactions are light-emitting, and this chemiluminescence may be used to quantify the concentration of some specific analytes. The major advantage of this approach is that no light source is required. Some commercial sensors based on this approach are already available; for instance, in the particular case of water quality monitoring, toxicity sensors based on this technique have been developed by a number of companies. Given that the light emission in chemiluminescence reactions is generally weak and emitted over a broad reaction area, the high gain and large active area of PMTs make them particularly suitable for this application. The use of compact battery-powered PMTs facilitates the development of handheld units.

5.4.6. Fiber Configurations

When using optical fibers for any of the spectroscopic measurements treated above, a key design issue is the optical arrangement and, in particular, the choice and number of optical fibers used. In general terms, the larger the fiber core cross-sectional area and numerical aperture NA, the more efficiently light can be coupled in to the fiber. An extreme example of this feature is the very low efficiency achievable when launching from a nonlaser source into a single-mode fiber. When dealing with absorption–transmission measurements, it is sensible to ensure that the light beam traversing the sample is parallel. This can be achieved by locating a small lens at its focal distance from the fiber tip. A more attractive option is to employ optical fibers that are preterminated with integral collimators such as GRIN/SELFOC lenses. In the case of fluorescence measurements, one or more optical fibers may be used. The key issues here are the separation of excitation and emission signals and the fluorescence collection efficiency. For example, if the same optical fiber is used for both excitation and fluorescence collection, then a method is required to enable efficient spatial separation of the returning signal and the launched radiation. Such methods include simple beamsplitters, dichroic beamsplitters, and, in the case of laser sources, mirrors with a small hole drilled through the center. In this last example, the laser excitation light is injected into the fiber through the hole in the mirror, which is oriented at 45° to the laser beam. The returning fluorescence signal will usually occupy the full NA of the fiber and will be emitted as a cone of light that gets deflected by the mirror to the detector, with minimal loss of signal through the hole. The issue of collection efficiency has attracted considerable attention in the context of sensors based on either fluorescence or Raman scattering. In the latter case, the method of choice is six collection fibers oriented symmetrically around the central excitation fiber. The outer fibers may be polished so that the end-face is not parallel to the fiber axis. This feature may be used to orient the fiber collection cone in the most favourable direction to enhance efficiency.

5.5.   EVANESCENT WAVE SENSORS

Most optical chemical sensors rely on conventional geometric configurations, whereby light is incident directly on a miniature cell or on a layer of immobilised reagent located in free space or at the distal end of an optical fiber. Significant advantages can be obtained, however, by adopting an alternative approach based on evanescent wave interactions. The origin of the evanescent wave may be explained as follows. When light propagates in an optical fiber or waveguide, a fraction of the radiation extends a short distance from the guiding region into the medium of lower refractive index that surrounds it. This evanescent field, which decays exponentially with distance from the waveguide interface, defines a short-range sensing volume within which the evanescent energy may interact with molecular species. Optical waveguide sensors for chemical and biological species based on such evanescent wave (EW) interactions have attracted considerable research interest [13]. The motivation for adopting the EW approach derives from a number of advantages offered by the technique in particular applications:

1. Because the interrogating light remains guided, no coupling optics are required in the sensor region and an all-fiber approach is feasible. Furthermore, considerable miniaturization is possible and this is particularly rele-

vant to integrated optic devices for which EW interactions are the predominant sensing mechanism.

2. By controlling the launch optics it is possible to confine the evanescent field to a short distance from the guiding interface and thereby discriminate to a large extent between surface and bulk effects (see Section 5.5.1). This is particularly important in some applications that involve surface interactions, such as fluoroimmunoassay [14].

3. The technique can provide enhanced sensitivity over conventional bulk-optic approaches. For example, fiber-based EW absorption devices are more sensitive than bulk-optic ATR (attenuated total reflection) crystals by virtue of the greater number of reflections per unit length (or, equivalently, the greater power in the evanescent field).

4. It is often difficult or inconvenient to perform accurate absorption measurements on highly absorbing or highly scattering media. Fiber-optic EW spectroscopy is suitable for such samples because the effective path length is so small and the technique is much less sensitive to scattering.

5. If an optical fiber is configured to be sensitive to EW interactions all along its length or at discrete zones, then fully- or quasi-distributed sensing is possible. This would enable monitoring of the spatial profile of an analyte concentration over substantial distances. Similarly, the line average of a species concentration could be acquired.

6. In contrast with conventional distal-face optrodes, the EW approach affords the sensor designer greater control over interaction parameters such as interaction length, sensing volume and response time.

EW sensors are not without difficulties. Chief among these is the problem of surface fouling, which, if significant, can reduce sensitivity and necessitate frequent recalibration until the sensor is no longer viable. A number of compensation techniques have been proposed, but these have not been implemented experimentally [15]. If suitable techniques are not found, then commercial EW devices will be restricted, in some applications, to short-term or disposable use.

5.5.1. Theoretical Background

An awareness of the critical parameters that determine the extent of EW interactions is important for the optimal design of EW sensors. Control over the degree of penetration of the evanescent wave into the low-index medium is important in some applications. This quantity is often characterized by the penetration depth d_p, which is the perpendicular distance from the interface at which the electric field amplitude, E has fallen to $1/e$ of its value, E_o, at the interface:

$$E = E_0 \exp\left(\frac{-z}{d_p}\right) \qquad (5.12)$$

The magnitude of the penetration depth is given by

$$d_p = \frac{\lambda}{2\pi n_1 [\sin^2 \Theta - (n_2 / n_1)^2]^{1/2}} \qquad (5.13)$$

where λ is the vacuum wavelength, Θ is the angle of incidence to the normal at the interface, and n_1, n_2 are the refractive index values of the dense and rare media, respectively. It is useful to note that n_2/n_1 in the denominator may be replaced by $\sin \Theta_c$. Although d_p is typically less than λ, it is clear from Eq. (5.13) that its value rises sharply as the angle of incidence approaches the critical angle Θ_c. This equation highlights the importance of the interface angle Θ in the design of EW sensors.

When considering absorption of the evanescent wave, the quantity of evanescent power that can interact with the analyte is a critical parameter. In the case of optical fiber EW sensors this quantity is closely related to the fraction r of the total guided power that resides in the cladding region:

$$r = \frac{P_{clad}}{P_{tot}} \qquad (5.14)$$

This fraction is determined to a large extent by the fiber V parameter, which was introduced in Section 5.3.1, and is given by

$$V = \frac{2\pi}{\lambda} a \, \text{NA} \qquad (5.15)$$

where a = fiber core radius, numerical aperture NA = $(n_1^2 - n_2^2)^{1/2}$, n_1 = core refractive index, and n_2 = cladding (analyte) refractive index. The dependence of r on V for individual modes in optical fibers is well known [16], and it can be shown that

1. Substantial values ($>50\%$) for r can be achieved in single-mode ($V < 2.405$) fibers.
2. Values of r are maximized for modes close to cutoff (which, in ray-optics terminology, are higher-order modes).
3. For high V-number values ($V \gg 1$), the average fractional power in the cladding is very low.

Although the high sensitivity achievable with single-mode fibers is very useful in some applications, highly multimode fibers (V >> 1) are most frequently used in EW sensing because of their much higher power throughput and ease of handling. Hale and Payne [17] have shown that the average value of r is given by $\frac{4}{3}V$ for weakly guiding ($n_1 \approx n_2$) multimode fibers in which all modes are propagating. Although many EW sensors do not comply with the weakly guiding condition and are not mode-filled, this expression gives an indication of the fraction of guided power available for EW sensing. For example, it yields a value of $r = 0.0006$ when light of wavelength 500 nm is propagated in a typical silica ($n_1 = 1.46$) multimode fiber core ($a = 300\ \mu m$) located in an aqueous ($n_2 = 1.33$) environment. Values of r less than 1% are typical in multimode EW sensors. In the case of sensors based on evanescent wave absorption one may write a modified version of Eq. (5.4):

$$I_T = I_0 \exp -(r\alpha L) \tag{5.16}$$

This expression, which is applicable in well-defined circumstances [18], describes the intensity transmitted through an EW absorption region of length L, where α is the absorption coefficient of the absorbing species and I_0 is the intensity transmitted in the absence of absorbing species. From Eq. (5.16) one can establish an expression for the evanescent wave absorbance:

$$A_{EW} = \log_{10}\frac{I_0}{I_T} = 0.434\,r\,\alpha\,L$$

It is clear from this equation that a critical feature in the design of EW absorption sensors is the optimization of the r value. The effective value of r can be increased by various techniques such as tapering or coiling the optical fiber, and higher-order mode selection via the use of spatial filters.

The principles of operation of EW fluorescence sensors are not so straightforward, however. If fluorescent species are located within the evanescent volume, a fraction of the fluorescence excited by the evanescent wave is coupled back into guided modes of the optical fiber. The efficiency of collection of this fluorescence by the fiber has been the subject of a number of theoretical and experimental studies [18,19]. Marcuse [20] has pointed out that fluorescent light in the cladding region couples to guided modes via the evanescent field tails of these modes. Consequently, modes near cutoff collect more light than do more strongly guided modes. Therefore the collected signal should increase roughly linearly with V for fluorescent material distributed throughout the cladding because the number of modes near cutoff increases linearly with V. It is clear, however, from the earlier parts of this section that the extent of evanescent wave absorption (characterized by r) is inversely proportional to V. Therefore EW excitation and collection of fluorescence exhibit opposing dependencies, and, as a result, the optimal design of such sensors is more complicated than for absorption sensors and generally requires a compromise approach. The theoretical considerations presented in this section identify many of the critical parameters involved in optimizing the performance of EW sensors that employ absorption or fluorescence. Practical implementation of these considerations generally involves careful selection of the launch and detection optics and the fiber dimensions.

5.6. DIRECT SPECTROSCOPIC SENSORS

Sensors that rely on the intrinsic spectral properties of the analyte require an appropriate light source with sufficient output power at the absorption or excitation band of the analyte together with a suitable detector. If, in addition, the sensor is to exploit optical fibers in order to carry out the measurement remotely, then the attenuation of the optical fiber in the relevant spectral region is a critical issue. This approach, sometimes referred to as *remote-fiber spectroscopy* (RFS), is routinely applicable in the vis–NIR spectral region, where high-quality telecom fibers transmit best, but more recently has benefited from the increased availability of mid-IR transmitting optical fibers (see Section 5.3.1). Although progress in improving the UV transmission of optical fibers has been much less significant, there have been a number of reports of sensors operating in this region. The near-IR and, especially, the mid-IR regions are richest in spectroscopic information, however, and consequently have attracted most attention, particularly in the areas of gas sensors and water monitors for hydrocarbon pollution.

A wide range of examples of RFS sensors has been reported in the literature, and extensive references to these can be found in the major reviews cited in Section 5.1. For the purposes of examining the RFS approach, we concentrate here on just three examples that highlight the principal features of the technique.

5.6.1. Nitrate Sensor Based on UV Absorption

A remote fiber optic sensor for measuring the nitrate concentration in river and well water has been demonstrated and shown to correlate well with standard ion chromatography method [21]. The nitrate ion represents a potential health hazard and its maximum admissible concentration (European Community Directive) is 11 mg/L NO_3–N (nitrate as nitrogen). The principle of operation of the sensor is based on the known ultraviolet absorption of the nitrate ion. The strong absorption of nitrate in the 210 nm region is measured relative to the absorption at a reference wavelength. The wavelength 275 nm is chosen for this purpose because nitrate does not absorb at this wavelength and it can be used to compensate for interference, mainly by organic species when these are present in significant concentra-

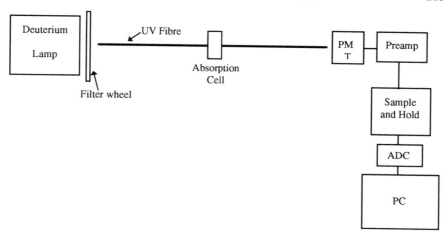

Figure 5.5. Schematic diagram of nitrate monitoring system.

tions. This dual-wavelength approach is achieved by using two narrowband filters centered respectively on 210 and 275 nm located on a rotating filter wheel placed between the deuterium lamp and the launch fiber as shown in Figure 5.5. This approach also compensates for spectrally neutral drifts within the system such as aging of the light source and detector. Dual-wavelength referencing is an important feature of almost all direct spectroscopic sensors.

UV-transmitting fibers are used to transmit light to and from a 1-mm pathlength transmission cell. The optical fiber used is Fibreguide Industries Superguide G with a core diameter of $600 \, \mu m$ and an attenuation of $1.5 \, dBm^{-1}$ at 210 nm. The detection system consists of a UV-sensitive photomultiplier tube and low-noise preamplifier. The output from the preamplifier is fed to an analog-to-digital converter via a "sample and hold" amplifier (SHA) that is synchronized to the rotating filter wheel. The SHA circuit effectively holds the signal at a constant value while an analog to digital conversion was carried out. Signals corresponding to light levels at 210 and 275 nm are stored by a computer. The detected signals are processed to yield an absorbance factor AF, which is defined as

$$AF = \log_{10} \frac{I_0}{I} \qquad (5.17)$$

The detected signal at 210 nm is represented by I and the signal at 275 nm represented by I_0. The spectral response of the system at 210 nm is very different from that at 275 nm. Consequently, when A is calculated using Eq. (5.17), the result is not a true absorbance expressible in absolute absorbance units. For this reason the calculated result was referred to as an *absorbance factor* (AF).

Measurements were carried out with this system over a total fiber length of 40 m, and the data exhibited a nonlinear calibration curve (absorbance factor vs. concentration), which resulted in a useful operational range of 0.4 to 30 mg/L NO_3–N. The limit of detection (LOD) of the sensor was found to be 0.4 mg/L NO_3–N and the repeatability in consecutive measurements was 1.7%. The sensor showed some susceptibility to interference (e.g., high concentrations of carbonates), although some of these effects can be reduced by restricting the ingress of specific interferents to the absorption cell by the use of an appropriate membrane. Further development of this system would allow it to become a more compact, portable instrument. For example, the use of a CCD array instead of the photomultiplier tube would allow the capture of complete spectra, thereby eliminating filters and moving parts from the design, with the added bonus of being able to compensate for turbidity and other interferences. Furthermore, this approach would facilitate the development of multiwavelength algorithms that would impart greater selectivity.

5.6.2. Hydrocarbon Gas Detection by Near-Infrared Absorption

Optical sensors for the detection of flammable, toxic, or otherwise relevant gases have attracted much interest. For example, methane has attracted particular attention because of the explosive hazard it presents in areas such as offshore rigs, coal mines, and landfill sites.

Methane (CH_4) is a highly flammable gas if its concentration is between 5% and 15% by volume in air. These two percentage limits are defined as the lower explosion limit (LEL) and the upper explosion limit (UEL), respectively. Above the UEL, there is not enough oxygen to support combustion. Research activity in this field has been reported since 1983, when Chan et al. first described a system with a limit of detection of 2000 ppm, based on absorption at the $v_2 + 2v_3$ combination band of methane at a wavelength of 1.33 μm, over a 2-km fiber length [22]. Subsequent work at this wavelength was reported by the same authors in 1987 when the system was used to remotely sense methane in city gas. An improved limit of detection of 1300 ppm was reported over 2 km [23]. Other groups have also developed systems at this wavelength [24,25].

In 1983, Hordvik et al. first reported a remote fiber-optic methane-sensing system operating at the $2v_3$ overtone of methane at 1.66 μm. The limit of detection of this system measured over a 600 m length of optical fiber was less than 5000 ppm of methane in nitrogen [26]. Working at this wavelength is attractive as the $2v_3$ overtone absorption is approximately twice as strong as the $v_2 + 2v_3$ combination band at 1.33 μm and is also transmitted well by silica fiber. Other research groups, including Chan et al. [27], Stueflotten et al. [28], Zientkiewicz [29], and more recently Dubaniewicz et al. [30], have demonstrated similar systems operating at 1.66 μm. Dubaniewicz et al. report a limit of detection of 2000 ppm over a 2-km length of fiber.

All the systems referred to above employ a "differential absorption" technique for sensing methane, whereby the signals corresponding to an absorbing and a non-absorbing (or reference wavelength) are measured alternately and ratioed. Because effects that are not dependent on the gas concentration (e.g., dust in the gas cell or source power fluctuations) are virtually independent of wavelength, taking a ratio of the two signals eliminates errors that may arise in concentration determination. The systems use interference filters or a monochromator to select the appropriate absorption and reference wavelengths.

Considerably enhanced sensor performance can be achieved, however, by the use of tunable DFB (distributed feedback) lasers [31]. Unlike conventional multi-mode diode lasers, which emit at a number of discrete wavelengths, DFB diode lasers emit light at a single wavelength and are capable of continuous wavelength tuning. In general, coarse control of the laser diode emission wavelength is achieved by adjustment of the laser temperature while fine-tuning is achieved by varying the injection current. Routine absorption measurements can be made by measuring the change in intensity of the laser beam transmitted through the gas as the laser emission wavelength is tuned across the absorption feature of interest. Such *direct detection* is useful at high concentrations of the gas. When dealing with very low gas concentrations, higher sensitivity is required, and this can be achieved by the use of *wavelength-modulation spectroscopy* (WMS or harmonic detection). Here the laser wavelength is set to a value close to the absorption feature of interest. The injection current of the laser is then slowly ramped to provide fine wavelength tuning across the absorption feature, while simultaneously imposing a sine-wave signal of frequency F (typically 1 kHz) on the DC current. As the laser wavelength scans the absorption line, the sine-wave modulation yields a signal at the second harmonic ($2F$) of the modulation frequency, and this may be synchronously detected, thereby providing high sensitivity and good signal-to-noise ratio (SNR). This technique has been demonstrated successfully for important gases such as methane [32], hydrogen sulfide [33], and CO_2 [33]. While it provides very low detection limits, the major drawbacks of the approach at this stage are the high cost of the DFB lasers (typically $1000s for single units) and the restricted range of wavelengths available.

5.6.3. Mid-IR Evanescent Wave Sensor for Chlorinated Hydrocarbons Based on Polymer Enrichment

A generic fiber-optic sensor, for the in situ detection of chlorinated hydrocarbons (CHCs) and some pesticides in water, has been developed by a number of groups, and one example is highlighted here [34]. The sensing element consists of a silver halide ($Ag Br_x Cl_{1-x}$) optical fiber coated with an appropriate polymer cladding, such as polyisobutylene (PIB) or Teflon. The polymer both enriches the chemical species to be measured in the evanescent wave region of the fiber and minimizes water interference. Evanescent wave spectrometry is then used to quantify the con-

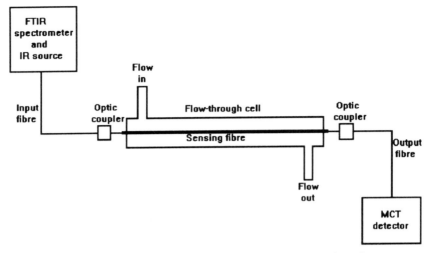

Figure 5.6. Schematic diagram of mid-IR fiber-optic evanescent wave absorption system.

centration of the enriched species such as CHCs that have their strongest absorption bands at wavelengths above 10 μm, where silver halide fibers transmit. The typical experimental configuration is shown in Figure 5.6. Using the fiber-optic interface accessory, infrared light from the FTIR spectrometer's blackbody source is launched into a midinfrared-transmitting (MIR) optical fiber. Light from this fiber is coupled into a 15-cm length of unclad 1000 μm MIR optical fiber that is mounted, using septa, in a flow-through sample cell. Light transmitted by the sensor fiber returns to a liquid nitrogen-cooled mercury–cadmium–telluride (MCT) detector. The unclad sensing fiber is first coated with the appropriate polymer and then mounted in the sample cell. The desirable properties for the polymer are high enrichment for a large range of analytes, hydrophobicity, fast response time, good reversibility, and a minimum number of absorption features in the spectral region of interest. Experiments show that PIB is a good polymer for enriching chlorinated hydrocarbons such as trichloroethylene (TCE), while pesticides show better enrichment in PVC/chloroparaffin coatings [35].

A typical evanescent wave absorbance spectrum measured with such a system is shown in Figure 5.7. Typical limits of detection for chlorinated hydrocarbons using short lengths of straight optical fibers are in the hundreds-parts-per-billion (100s-ppb) region. Higher sensitivity can be achieved, however, by enhancing the evanescent field penetration depth, by coiling or tapering the sensor fiber, and/or by using laser diode sources. Limits of detection close to 1 ppb have been achieved in this manner.

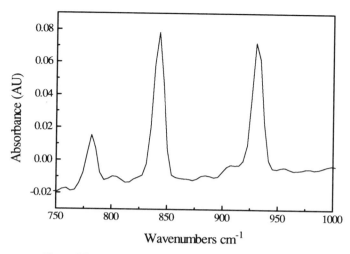

Figure 5.7. Evanescent wave absorption spectrum for TCE.

5.7. REAGENT-MEDIATED SENSORS (OPTRODES)

Most analytes of interest do not possess intrinsic spectral properties that facilitate
the direct spectroscopic sensing approach, especially where the use of low-cost optical sources and detectors is a requirement. In these circumstances, intermediate
reagents, which respond optically (e.g., by absorption or fluorescence change) to
the analyte, may be employed to provide the sensor transduction mechanism. Such
reagent-mediated sensors (optrodes) often provide greater sensitivity than do their
direct spectroscopic counterparts. Although a wide range of optical tests for analytes has been established over the years as analytical chemistry has developed,
many of these tests do not possess the important sensor characteristics of reversibility and selectivity. The optimal sensor reagent is very sensitive to the analyte, exhibits complete reversibility on removal of the analyte, responds quickly,
and shows a high degree of selectivity for the analyte. This last feature broadens
the applicability of the system and obviates the need for sample conditioning before exposure to the sensor. If the reagent system is not reversible, then the measurement device is conventionally termed a *probe* as opposed to a sensor. Probes
may be used successfully for continuous monitoring over defined time periods if
operated in a renewable reagent mode or if the measured parameter is the rate of
change of colour formation, for example. The vast majority of reagent-based sensing systems, however, are based on reversible indicators and this section deals only
with these. The ideal optrode reagent system is one that is compatible with low-

cost semiconductor light sources (LEDs, LDs) and detectors, exhibits photostability, does not suffer from reagent leaching, and has the desirable reagent properties listed earlier in this section.

5.7.1. Reagent Immobilization

Whether the optrode employs an optical fiber, waveguide, or neither, a key design issue in sensor fabrication is the immobilization of the reagent molecules in an appropriate matrix. A number of immobilization methods may be used, but, in any particular circumstance, the choice of method should be based on the following considerations:

1. The nature of the reagent and its compatibility (e.g., with respect to size or charge) with the immobilization method.
2. The degree of concern over leaching; for example, leaching cannot be tolerated when in vivo biomedical sensing is employed.
3. The influence of the immobilization method on the spectral and/or sensing properties of the indicator.
4. The complexity of the immobilization method and its reproducibility.
5. The requirement for particular physicochemical properties of the support matrix; such properties could include hydrophobicity or high permeability for a particular analyte.

An additional consideration, which is crucial for the eventual commercialization of a sensor, but which is often neglected in laboratory-based demonstrations, is the ease with which the immobilization method can be transferred to mass production in an industrial environment.

Electrostatic binding may be used for indicators that remain charged over the full measurement range of interest. This process is generally used with ion-exchange resins. Adsorption onto polymeric substrates or microspheres has been used widely in sensors but is prone to leaching problems. Covalent binding is the most reliable method for indicator immobilization and by nature excludes the problem of leaching. It is, however, the most complex of the immobilization processes, often requiring chemical modification of both the indicator and the support matrix in order to facilitate the binding reaction. Binding can take place directly on the optical substrate (optical fiber or waveguide) or more usually on the support matrix. In the case of silica-glass materials, for example, surface silanization is generally required in order to activate the surface for binding.

Polymer films are often used to entrap reagents but can exhibit serious leaching problems. These difficulties can be alleviated by chemically binding the small reagent molecule to the polymer. By using hydrogels such as polyacrylamide, poly(vinyl alcohol), or hydrolyzed cellulose acetate, sensors suitable for detection in aqueous solution can be developed. These hydrogels swell when in contact with water, which increases the rate at which the sample components diffuse into the coating. In this way, the response time of the sensor is reduced.

Alternatively, films can be prepared by copolymerizing a monomer with a suitable polymerizable derivative of the optical reagent to form a polymer film that contains a high concentration of the reagent that is irreversibly bound within the matrix. The diversity of monomers, and binding chemistries that can be used to attach optical reagents to them, gives rise to a wide range of optical reagents. However, it is important to note that the electronic state energies and lifetimes of these chemical reagents may be altered when bound within a polymeric matrix. Therefore, basic characterization of functionalized polymer films is an integral part of research in this area.

The physical entrapment of reagents in a porous glass material fabricated by the sol-gel process has been attracting increasing attention in recent years because of the design flexibility afforded by the process, especially in the preparation of silica-based materials [36]. These materials offer superior properties of stability and durability in comparison with polymers. Although leaching may still be an issue here, the approaches listed above for polymers may also be applied.

5.7.2. Examples of Optrodes

As was the case for the direct spectroscopic sensors treated earlier, it is not the intention here to provide a comprehensive literature review of reagent-mediated sensors. Rather, examples of a number of important approaches are presented and these highlight the immobilization methods, sensor configurations and transduction mechanisms employed.

5.7.2.1. pH Sensor for Blood Analysis

Because of its importance in biomedical, industrial, and environmental processes, reports of sensors for pH considerably exceed all other optical chemical sensors in the literature. pH optrodes are based on pH-dependent changes in the optical properties of the indicator phase, which reacts reversibly with protons in the sample. The most popular designs use the pH-dependent absorption or fluorescence of optical reagents that are weak electrolytes and exist in both acidic and basic forms over the pH range of interest. The acid and corresponding conjugate base participating in the pH-dependent equilibrium are selected to have different absorption or fluorescent properties. The equilibrium may be represented by

$$HA \leftrightarrow H^+ + A^-$$

where HA denotes the acidic form of the dye and A^- the basic form. Since the degree of dissociation depends on the pH, the acidity level can be determined by measuring the relative concentrations of both forms of the dye. The issue of what optical pH sensors actually measure has been the subject of some debate and is treated comprehensively in an important paper by Janata [37]. In this work, he points out that most reports of pH optrodes ignore the effect of ionic strength on the dissociation equilibrium of the indicator. This omission is acceptable in very dilute solutions but can lead to serious errors in real-world environments, where the ionic strength may vary to a significant extent.

A commercially available blood pH sensor is chosen here to represent a successful approach to optical pH sensing. It is part of system developed by AVL List, Austria for single-shot in vitro blood gas and pH analysis [38]. The pH optrode is based on the pH-dependent changes of the fluorescent indicator 1-hydroxy-pyrene-3,6,8-trisulfonate (HPTS), which is suited to the physiologically important measurement range pH 7.0–7.5. A cross section of the generic planar optrode material is shown in Figure 5.8. The sensing layer contains the immobilized fluorescent dye and is attached on top of a transparent polyester foil. The sensing layer is covered with a black optical isolation layer, permeable to the analyte. The optical isolation layer separates the sensing layer from optical interferences due to calibrants or the sample. The bottom part of the foil is covered with an optically transmissive adhesive layer. The optrode is manufactured as a compound foil. Individual sensor disks (3 mm in diameter) are punched out and placed into disposable measuring cells.

For preparation of the pH-sensitive layer, aminoethylcellulose fibers (diameter 8 μm) are reacted with 1-hydroxy-pyrene-3,6,8-trisulfochloride. A substantial fraction of amino groups of the cellulose material remain unreacted. The remaining amino groups are reacted with acetic anhydride to prevent formation of high local charges due to formation or ammonium ions at neutral pH. The pH-sensitive layer is obtained by attaching the dye-loaded cellulose fibers to the surface of a polyester foil (thickness 95 μm) and embedding them in an ion-permeable polyurethane-based hydrogel material. For optical isolation of the pH-sensitive compound foil, a polyurethane-based hydrogel layer containing carbon black pigments is cast on the pH-sensitive layer.

With this setup, signal-to-noise ratios ranging from 2000 to 4000 can be achieved. When calibrated, measurements made on human blood samples with the

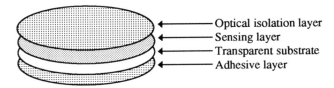

Figure 5.8. Cross section of pH sensor.

optrode-based measuring system demonstrated calibration and measurement reproducibility comparable to that of a state-of-the-art blood pH analyzer based on
a pH electrode [38]. It should be noted that a variant of the optrode just described
can function as a CO_2-sensitive optrode if minor modifications are made to the
fabrication. The only differences are

1. The pH-sensitive layer is equilibrated with a 0.022 M $NaHCO_3$ solution.
2. The sensing layer is covered with a CO_2 permeable, ion-impermeable black
   silicone layer.

The combined use of a bicarbonate buffer with a pH sensor is the conventional approach to fabricating CO_2 optrodes.

5.7.2.2. Fluorescence-Based Oxygen Sensor

The determination of oxygen concentration is of major importance in many industrial, medical, and environmental applications. Optical oxygen sensors are
more attractive than conventional amperometric devices because they do not con-
sume oxygen, and are not easily poisoned. Sensor operation is usually based on
the quenching of fluorescence in the presence of oxygen as described by the
Stern–Volmer equations (see Section 5.4.3).

Luminescent transition-metal complexes, especially ruthenium poly(pyridyl)
compounds, are particularly attractive as oxygen-sensing species. The ruthenium
complex [Ru^{II}-tris(4.7-diphenyl-1,10-phenanthroline)] is widely used as an oxygen-sensitive indicator because of its highly emissive metal-to-ligand charge-
transfer state, long lifetime, large Stokes shift, and strong absorption in the blue-
green region of the spectrum. Oxygen sensors based on the immobilization of this
ruthenium complex in nanoporous silica films prepared via the sol-gel process
have been reported [39], and the work is summarized here.

Sol-gel-derived silica films are fabricated from silicon alkoxide precursors that
undergo hydrolysis and polycondensation reactions, followed by a temperature
program that controls the densification process. If the temperature is limited to less
than 200°C, a nondensified porous glass matrix is formed. This nanoporous glass
acts as a support matrix for analyte-sensitive dyes that are added to the silicon
alkoxide solution. The dye molecules are entrapped in the nanometre-scale cages
formed by the crosslinking silicon and oxygen units. Smaller analyte molecules
can permeate the matrix and access the dye complex in the pores. With selection
of fabrication parameters appropriate to the size of the dopant molecule, leaching
is negligible.

The standard film fabrication process involves mixing a silicon alkoxide precursor with water, ethanol, and hydrochloric acid, which acts as the catalyst. The
precursors normally used are either tetraethoxysilane (TEOS), or organically mod-

ified derivatives such as methyltriethoxysilane (MTEOS). Typically the water is kept at pH = 1 and the R value, the ratio of water to silicon alkoxide precursor, at 4. The ruthenium complex is added to the precursor solution which is stirred for 1 h. The typical concentration of ruthenium complex used is 2.5 g/L with respect to the precursor solution. After stirring, the sol is stored at 70°C for 18 h to promote hydrolysis and condensation polymerization. The sol is then dip-coated in a draught-free environment onto planar substrates using a computer-controlled dip-coating apparatus. A typical coating speed is 1 mm/s, which gives a film thickness of about 300 nm. Dip-coated films are then dried at 70°C for 18 h.

A typical measurement system is shown in Figure 5.9. The excitation source is a high-brightness blue LED whose spectral output peaks at 450 nm, and that has good overlap with the absorption spectrum of the ruthenium complex. The excitation light passes through a band-pass filter (bandwidth = 400–505 nm, centered at 450 nm) before impinging on the coated substrate, which is held at 45° to the excitation beam in the sample chamber. The fluorescence from the coated substrate passes through a long-wave pass filter (λ_{cuton} = 550 nm), and is focused onto a silicon photodiode detector. The filter combination minimizes the detection of the excitation light with minimal reduction of the fluorescence signal. This all-solid-state system serves not only to provide a high signal-to-noise ratio, but also allows for miniaturization and facilitates portability.

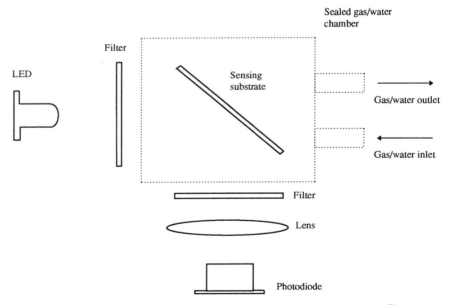

Figure 5.9. Experimental system for characterization of sol-gel oxygen-sensing films.

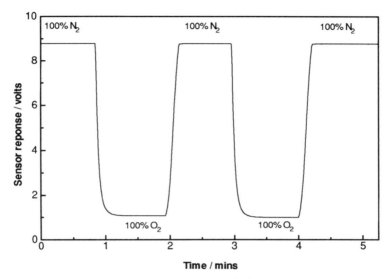

Figure 5.10. Sensor response to alternate environments of 100% oxygen and 100% nitrogen for a TEOS-based sol-gel film.

The response of a TEOS-based film to alternate environments of 100% oxygen and 100% nitrogen gas is shown in Figure 5.10. The response is characterized by a short response time, a high signal-to-noise ratio, and good reversibility. Moreover, the overall percentage quenching of the fluorescence signal between the two extremes of measurement is almost 90%. However, if this same film is exposed to alternate environments of fully oxygenated and fully nitrogenated (deoxygenated) water, the percentage quenching is only 20% as shown in Figure 5.11. The origin of the low quenching response is a combination of the low oxygen concentration in water and the hydrophilic nature of the TEOS sol-gel film surface. TEOS-based films have a high surface coverage of silanol ($Si-OH$) groups, which facilitate water adsorption on the surface of the film and hence the surface is hydrophilic. A hydrophobic film surface should enhance the dissolved oxygen (DO) quenching process by causing the partitioning of oxygen out of solution into the gas phase, within the sensing film. This can be achieved by the use of modified precursors of silica in the sol-gel film fabrication process. This serves to replace the majority of the surface silanol ($Si-OH$) groups with $Si-R$ groups, where $R = CH_3$, for example. These groups have an poor affinity for water, and thus the surface is rendered hydrophobic. Figure 5.12 shows the quenching response of a MTEOS:TEOS 3:1 film in aqueous phase, where the percentage quenching now has a value of 70%. Although the response time appears to be much longer than in gas phase, this is due mainly to instrumental considerations such as the volume of

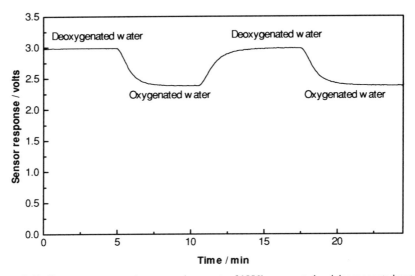

Figure 5.11. Sensor response to alternate environments of 100% oxygenated and deoxygenated water for TEOS films.

the sample cell and the sample delivery system. The major enhancement of sensor performance exhibited in Figure 5.12 highlights the versatility of the sol-gel process as a generic immobilization technique.

5.7.2.3. Calcium Ion Sensor

There is substantial interest in ion-selective optrodes that can sense clinically important alkali and alkaline-earth ions in both serum and whole blood. The example presented here deals with the Ca^{2+} ion but represents a generic approach that can be applied to a wide range of ionic analytes. The technique developed by the group of Simon at the ETH in Zurich was used initially for ion-selective electrodes but has since been modified for optrodes (this issue is discussed in greater detail in Chapter 2). The optrode is based on molecular recognition and ion exchange and consists of a polymeric membrane in contact with the sample solution. The active components are uniformly entrapped and homogeneously dissolved in the bulk of the polymer membrane (m), which is usually composed of plasticized poly(vinyl chloride) (PVC). The calcium-selective optrode membrane relies on the recognition and extraction of the calcium ion from the sample solution(s) into the membrane by a known calcium-selective ionophore [40]. Since the ionophore has no useful optical properties, a chromophoric pH indicator CH^+ is also incorporated in the sensor membrane to provide optical transduction of the recognition process. Since electroneutrality within the sensing layer must be maintained, H^+

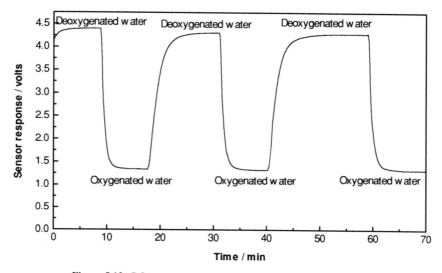

Figure 5.12. DO sensor response of MTEOS:TEOS 3:1 sol-gel film.

ions attached to the protonated basic indicator are exchanged and released into the sample:

$$Ca^{2+} + L + 2CH^+ \rightarrow 2H^+ + CaL^{2+} + 2C$$
$$\text{(s)} \quad \text{(m)} \quad \text{(m)} \quad \text{(s)} \quad \text{(m)} \quad \text{(m)}$$

The particular indicator used exhibits this deprotonation as a reduction in absorption at $\lambda = 660$ nm, which is easily measured using a red LED. Clearly the sample solution needs to be buffered at a pH value that is far from the sensitive region of the indicator in order to avoid indicator absorbance changes that are not related to the analyte. When this is carried out, the measured optical absorption may be used to derive the calcium ion concentration. All the materials required for the above sensor, as well as ionophores for other ions, are available from Fluka AG, Switzerland. The approach described here is clearly generic and has the potential for widespread application to the sensing of ions.

5.8. SIGNIFICANT DEVELOPMENTS

Among the vast range of optical chemical sensors reported in the literature in the past 15 years, a small number of key approaches has emerged. These approaches are most likely, in the author's opinion, to make a significant impact on commercial chemical sensing technology. In addition to the examples already presented in sections 5.6 and 5.7, these key areas are as described in the following paragraphs.

5.8.1. Submicrometer-Scale FOCS

A combination of tapered optical fibers, near field optics and photochemical synthesis has been used to produce submicrometer-scale optical chemical sensors that have been applied successfully in chemical measurements in micrometer domains such as single biological cells. Fluorescent dye-doped polymers are photopolymerized in the near-field region at the tip of a metal-coated fiber-optic taper. This produces a sensor with a very short response time (milliseconds or less), a very high spatial resolution, and a very low sample volume requirement. The major work in this area has been performed by the group of Kopelman at the University of Michigan [41]. Typically, single-mode fibers are heated (e.g., by a CO_2 laser) and drawn by a micropipette puller to produce a taper with a tip diameter of 0.1–0.5 μm. The sides of the fiber tip are then coated with aluminum, leaving the end-face as a transmissive aperture. Immobilization of the sensing reagent is achieved by first silanizing the fiber tip and then immersing it in the polymerization solution containing the dye. Photopolymerization is then stimulated locally at the fiber tip by argon-ion laser radiation transmitted by the optical fiber.

While this configuration has many advantages, it is suited to laboratory use only and requires expensive and bulky associated instrumentation. The fiber tip generates a fluorescence signal with high spatial resolution, but this emission must be collected by a modified microscope linked to a sensitive photodetection system, such as a photon-counting PMT. This technique has been applied successfully to the sensing of a range of analytes including oxygen, pH, and glucose.

5.8.2. Fluorescence Lifetime-Based Sensing and Imaging

Although most types of fluorescence sensing and imaging are based on measurement of fluorescence intensity, this approach is susceptible to a number of problems, including photobleaching of the dye, leaching, fouling, and source/detector aging. All of these may result in a change of fluorescence signal that is unrelated to the analyte concentration. Although ratiometric measurements based on dual-wavelength referencing can be used to compensate somewhat for these problems, this adds to the complexity and cost of the sensor. Measurement methods that are based on the fluorescence lifetime of an indicator, however, offer a solution to these problems. The fluorescence lifetime is an intrinsic property of the indicator molecule, is generally insensitive to intensity fluctuations, and provides an inherently referenced signal. A range of quenching and molecular interactions result in changes in the lifetime of fluorophores and these have been exploited in sensor systems [42,43]. The most commonly used measurement method is phase fluorometry, in which the fluorophore is excited by sinusoidally modulated light and the lifetime-dependent phase shift in the emitted light is detected. An additional exciting development is fluorescence lifetime imaging microscopy (FLIM) which enables the recording of images where the contrast is based on a 2D fluorescence lifetime distribution [43].

5.8.3. Fluoroimmunosensors

The combination of the high specificity of antibody recognition with evanescent wave sensing has resulted in a range of high-performance fluoroimmunosensors [44]. In such sensors the antibodies immobilized on the surface of a fiber core bind fluorophore-labeled antigens in the evanescent wave region of the optical fiber. The characteristic penetration depth of the evanescent wave provides a spatial separation that discriminates between fluorophores bound to the core and those in free solution, thereby eliminating the washing steps required in other types of immunoassay. A crucial step in the development of these sensor systems was the optimization of the geometry of the fiber substrate to maximize the evanescent wave fluorescence collection efficiency. The optimal design was identified as an optical fiber with two tapered zones—the so-called combination taper. An additional important aspect of this work has been the use of long-wavelength fluorophores (λ_{exc} > 600 nm) for labeling proteins. Typical labels are Cy5 and Cy5.5, which are available from Amersham. This approach reduces the problem of background natural fluorescence, facilitates the use of low-cost laser diodes, and thereby accelerates the development of portable biosensor systems. The key work in this area has been performed by the group of Ligler at the Naval Research Laboratory, Washington DC, USA [45].

5.8.4. Surface Plasmon Resonance Sensors

Sensor systems based on surface plasmon resonance (SPR) have already made a commercial impact and are likely to continue to do so. A *surface plasmon* is a collective oscillation of free electrons in a metal film. The plasmon can be excited under strict conditions by the evanescent wave of totally internally reflected light. Resonance results in sharp minimum in the reflected light at a precise angle of incidence. The conditions for excitation are very sensitive to the refractive index of the medium on the outer surface of the metal film, and this sensitivity has been exploited in gas-sensing and biosensing applications [56]. Essentially, the system functions as a highly sensitive refractometer with selectivity imparted by the chemical–biochemical material coated on the sensing side of the metal film (usually silver or gold). For example, antibody–antigen reactions on the surface of the metal change the dielectric constant and result in a shift in the SPR angle. The sensitivity of a SPR system is directly related to the accuracy with which this angular shift can be measured. Moreover, sensitivity will be higher for larger analyte molecules because their effect on the refractive index will be proportionately greater. The major problem for these systems, however, is the susceptibility to nonspecific binding or adsorption, which will also result in a SPR angle shift.

A number of configurations ranging from bulk prism optics to planar waveguides can be used. The BIAcore system by Pharmacia, Sweden is now well established and offers impressive performance as a biosensor development system, enabling real-time monitoring of binding kinetics. The system employs disposable gold-coated sensor slides that can be used for multiple measurements. A fiber-optic probe configuration has been commercialized recently by the same company. Texas Instruments, USA have also developed a compact SPR probe with fully integrated optoelectronics in a prismatic head of centimeter dimensions.

5.8.5. Multianalyte Imaging Systems

There is considerable interest in the development of sensor systems capable of simultaneous detection of a range of chemical or biological analytes. Although several individual sensors could be multiplexed to achieve this goal, there is great benefit in exploiting advanced imaging technology to interrogate a single substrate on which multiple sensing sites are deposited. CCD (charge-coupled-device) cameras with a range of selectable operational features (cooled/uncooled, image intensifiers, pixel size) are routinely available for this purpose.

Walt et al. [47] have developed an innovative approach to multiple-analyte chemical sensing. The key element in the system is an imaging fiber bundle, which comprises several thousand individual fibers drawn together to produce a single solid unit of submillimeter diameter. Alternatively, discrete optical fibers may be bundled together as required. In either case, discrete sensing regions are created at the distal end of the fiber bundle by site-selective photopolymerization of monomers (doped with different analyte-sensitive fluorescent dyes), using the distinct optical pathways of small regions of the imaging bundle or the individual fibers. The bundle is then used to guide suitable excitation light from a single source to the multisensor head. Using a CCD video acquisition system, information can be collected at video rates (every 33 ms) for each of the different regions of the multifiber array. Simultaneous monitoring of pH, CO_2 and O_2, using a single imaging fiber bundle of 0.35 mm diameter, has been demonstrated with this technique. The technique has also been used to develop a vapor-sensing device, where the sensor array provided the input to a pattern recognition system incorporating artificial neural networks [48].

5.8.6. Integrated-Optic Devices

The planar waveguide offers a very attractive option as a sensor configuration, combining the advantages of thin films and evanescent wave sensing in a geometry that is convenient to manufacture and handle. Furthermore, it lends itself to the incorporation of additional optical features (e.g., multiple channels, splitters, diffractive optical elements) in so-called integrated-optic devices. Planar waveguides may be fabricated using a range of techniques and materials, including injection molding of polymers and dip-coating of sol-gel films. The basic system consists of a waveguiding layer on a planar substrate. The sensor layer is a reagent-doped coating of lower refractive index, coated on top of the waveguide layer. A critical design issue is the method used to launch light into the waveguide. End-fire coupling and injection through surface-relief diffraction gratings are the most attractive options.

Some new waveguide configurations are emerging, and these seek to integrate several sensor components in a chiplike unit. The advanced ". . . sensor on a chip" concept combines light sources, detectors, waveguides, and coupling optics in a single compact package compatible with a printed-circuit board. This approach, pioneered by Texas Instruments, is generic in that it may act as a platform for sensors based on fluorescence, absorption, or refractive index. The approach also lends itself to mass production and reduces assembly requirements.

5.9. LIMITING FACTORS

It is clear that optical chemical sensing technology offers great potential for the development of novel measurement systems for use in a wide range of important application areas. In the case of reagent-mediated sensors, in particular, this potential has not yet been realized. This is due to a number of factors, most of which are associated with the long-term calibration stability of the sensor. These factors include

Leaching of reagents
Stability of the immobilization matrix
Susceptibility to interferents and fouling

The sensor developer must pay particular attention to these issues in terms of both sensor design and characterization. Otherwise, the field of optical chemical sensing will continue to report interesting laboratory-based, proof-of-principle demonstrations of sensor potential but fail to realize its clear potential.

5.10. CONCLUSIONS

Optical chemical sensing technology is still in the development stage and continues to exploit advances in areas such as optoelectronics, materials science, and synthetic chemistry. Given the exciting developments that are emerging, it is likely to make a major impact on measurement technology in industrial, environmental, and biomedical applications in future years.

REFERENCES

1. Wolfbeis,, O.S., ed, *Fiber Optic Chemical Sensors and Biosensors,* Vol. I/II, CRC Press, Boca Raton, FL, 1991.
2. *Sensors Actuators* **B11** (1993); *Sensors Actuators* **B29** (1995).
3. See e.g., *SPIE Proc.* **2068** (1993).
4. Taib, M. N., and Narayanaswamy, R., *Analyst* **120,** 1617 (1995).
5. Norris, J. O. W., *Analyst* **114,** 1359 (1989).
6. Chudyk, W., et al., *Anal. Chem.,* **57,** 1237, (1985).
7. Sensors Unlimited Inc., Princeton, N.J., USA.
8. McAleavey, F. J., and MacCraith, B. D., *Electron. Lett.* **31**(10), 800 (1995).
9. MacCraith,, B. D., O'Keeffe, G., McDonagh, C., and McEvoy, A. K., *Electron. Lett* **30**(11), 888 (1994).

Lett. **29**(19), 1719 (1993).

:o-Technical Institute, 26 Polytechnich-

st Rd., Dunedin, FL, USA.

(1993), and references cited therein.

;lovacek, R. E., *Biosensors Bioelectron.*

onovic, I., *SPIE Proc.* **1314,** 262 (1990).

on. **8** (5,6), 743 (1993).

(1988).

, 237 (1992).

; (1988).

erty, A. P., MacCraith, B. D., Diamond,

994).

tt. **43,** 634 (1983).

vave Technol. **LT-5,** 1706 (1987).

24. King, T. A., and Mohebati, *SPIE Proc.* **1011,** 183 (1988).
25. Zientkiewicz, J. K., *SPIE Proc.* **992,** 182 (1988).
26. Hordvik, A., Berg, A., and Thingbo, D., *Proc. 9th Eur. Conf. Optical Communication,* 1983, p. 317.
27. Chan, K., Ito, H., and Inaba, H., *Appl. Opt.* **23,** 3415 (1984).
28. Stueflotten, S., et al., *SPIE Proc.* **514,** 87 (1984).

29. Zientkiewicz, J. K., *SPIE Proc.* **1085,** 495 (1989).
30. Dubaniewicz, T. H., and Chilton, J. E., *Report on Investigation G407,* U.S. Dept. Interior, Bureau of Mines, 1991.
31. Tanbun-Ek, T., et al., *J. Cryst. Growth* **107,** 751 (1991).
32. Weldon, V., Phelan, P., and Hegarty, J., *Electron. Lett.* **29,** 561 (1993).
33. Weldon, V., O'Gorman, J., Phelan, P., Hegarty, J., and Tanbun-EK, T., *Sensors Actuators* **B29,** 101 (1995).
34. Walsh, J. E., MacCraith, B. D., Meaney, M., Vos, J. B., Regan, F., Lania, A., and Arjushenko, S., *SPIE Proc.* **2508,** 233 (1995).
35. Regan, F., Meaney, M., Vos, J. G., MacCraith, B. D., and Walsh, J. E., *Anal Chim. Acta* **334,** 85 (1996).
36. MacCraith, B. D., McDonagh, C., O'Keeffe, G., McEvoy, A. K., Butler, T., and Sheridan, F. R., *Sensors Actuators* **B29,** 51 (1995).
37. Janata, J., *Anal. Chem.* **59,** 1351 (1987).
38. Leiner, M. J. P., *Sensors Actuators* **B29,** 169 (1995).
39. MacCraith, B. D., O'Keeffe, G., McEvoy, A. K., McDonagh, C., McGlip, J. F., and O'Kelly, B., *Opt. Eng.* **33**(12), 3861 (1994).
40. Spichiger, U. E., Citterio, D., and Bott, M., *SPIE Proc.* **2508,** 179 (1995).
41. Tan, W., Shi, Z. Y., and Kopelman, R., *Anal. Chem.* **64,** 2985 (1992).
42. Szmacinski, H., and Lakowicz, J. R., *Sensors Actuators* **B29,** 16 (1995).
43. O'Keeffe, G., MacCraith, B. D., McEvoy, A. K., McDonagh, C., and McGilp, J. F., *Sensors Actuators* **B29,** 226 (1995).
44. Wise, D. L., and Wingard, L. B., eds., *Biosensors with Fiber Optics,* Humana Press, Clifton, NJ, 1991.
45. Shriver-Lake, L. C., Golden, J. P., Patonay, G., Narayonon, N., and Ligler, F., *Sensors Actuators* **B29,** 25 (1995).
46. Liedberg, B., Lundstrom, I., and Stenberg, E., *Sensors Actuators* **B11,** 63 (1993).
47. Ferguson, J. A., Healey, B. G., Bronk, K. S., Barnard, S. M., and Walt, D. R., *Anal. Chim. Acta* **340,** 123 (1997).
48. White, J., Kauer, J. S., Dickinson, T. A., and Walt, D. R., *Anal. Chem.* **68,** 2191 (1996).

CHAPTER

6

MINIATURIZED CHEMICAL SENSORS

ROBERT J. FORSTER

School of Chemical Sciences, Dublin City University, Dublin Ireland

6.1. INTRODUCTION

Chemical sensors are devices that provide information about the types, concentrations, and chemical states of the species present within a sample. The *ideal* sensor can be inserted into the sample and will display the result of the chemical analysis within a few seconds with high precision and selectivity. No sampling, dilution, or reagent addition is required, and changes in the analyte concentration or activity can be displayed in real time. In this chapter, we focus on sensors whose transduction mechanism is based on mass or current changes occurring within the sensor when the analyte of interest is present. We place particular emphasis on miniaturized sensors and highlight recent developments in sensor arrays and chemometrical approaches to interpreting signals generated by sensor arrays. The variety of systems that can now be probed using chemical sensors is truly vast, ranging from the secretions of single cells, to the inside of reaction vessels run at high temperatures and pressures. Chemical sensors can routinely and accurately detect the concentrations of chemical species in the solid, liquid, or gas phases at concentrations as low as parts per trillion. As demonstrated by Table 6.1, there is a large market for chemical sensors ranging from continuous monitoring of chemical processes in industry, to carbon monoxide sensing in homes.

To operate successfully, the output from a chemical sensor should ideally depend only on the presence and concentration of the analyte of interest; thus, the sensor should be specific for a single target molecule. However, in reality, most sensors respond in a selective manner; that is, their response is dominated by a single chemical species, but they also exhibit minor responses to other species that act as interferents. As illustrated in Figure 6.1, many sensors achieve a selective response through a "lock and key mechanism" in which only the target molecule, or closely related analogs, fit the "lock" within the sensor to ultimately generate an electrical output.

Table 6.1. Typical Applications of Chemical Sensors

Application	Example
Automotive	"Intelligent" fuel management systems, emission monitoring
Aerospace	Systems monitoring, air quality sensing within cabin
Agriculture	pH detection, controlled application of herbicides and pesticides
Chemical industry	Materials testing, emission control, systems monitoring
Safety	Gas detection
Environmental monitoring	Detection of pollutants in air, water, and soil
Medicine	Determination of the concentration of anesthetic gases, clinical diagnosis
Customs	Detection of illegal and dangerous substances, drugs, and explosives
Quality control	Probing chemical composition, smell, freshness and flavor of foods, compliance checking of chemicals

6.2. MASS-SENSITIVE SENSORS

6.2.1. Introduction

That piezoelectric crystals are capable of very accurately measuring small changes in mass makes them especially attractive platforms on which to build chemical sensors. Selectivity toward a particular chemical species is achieved by depositing a chemically sensitive film over the mass-sensitive device.

In this section, we consider two types of mass-sensing phenomena, namely, bulk and surface acoustic wave devices. Both of these approaches employ piezoelectric crystals that are deformed mechanically by applying a potential or voltage in a controlled manner. By applying such a waveform, acoustic waves are generated that travel either through the bulk of the crystal, or along its surface. These devices are useful for chemical sensing because the velocity of the waves, and hence their frequency, depends on the mass of the crystal.

6.2.2. Bulk Acoustic Wave (BAW) Devices

Sharp resonant frequencies are obtained when an electrical stimulus is applied to a quartz crystal resonator. While the frequency is rather insensitive to changes in physical parameters such as temperature and pressure, it depends strongly on the mass of the oscillator. Therefore, since it is possible to measure the resonance frequency accurately using a frequency response analyzer, a sensitive chemical sen-

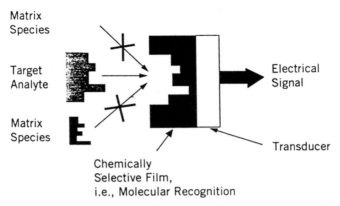

Matrix
Species

Target
Analyte

Matrix
Species

Electrical
Signal

Transducer

Chemically
Selective Film,
i.e., Molecular Recognition

Figure 6.1. Simplified model showing the origin of selectivity in sensor responses. The diagram illustrates docking of a single molecule type to a specific site located within a chemically sensitive film.

sor can be made by coating the crystal with a modifying film that selectively binds a target analyte. Table 6.2 details the wide variety of materials that can be used to fabricate mass-sensitive sensors and their corresponding acoustic impedances. For optimum performance, the acoustic impedance of the analyte and the piezoelectric crystal should be closely matched.

First, we consider the underlying principles on which BAW devices operate. When a voltage is applied to a piezoelectric crystal, several fundamental wave modes are obtained: longitudinal, lateral, and torsional, as well as harmonics. Depending on the way in which the crystal is cut, one of these principal modes will predominate. In practice, the high-frequency thickness shear mode is often chosen

Table 6.2. Acoustic Impedances of Materials Used to Fabricate Mass-Sensitive Devices

Material	Impedance 10^6 kg s m^{-2}	Material	Impedance 10^6 kg s m^{-2}
Platinum	36.1	Silver	16.7
Chromium	29.0	Silicon	12.4
Nickel	26.7	Indium	10.5
Aluminum oxide	24.6	Quartz	8.27
Palladium	24.6	Aluminum	8.22
Gold	23.2	Graphite	2.71
Copper	20.3		

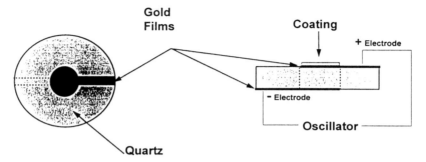

Figure 6.2. Quartz crystal microbalance configuration involving standing shear waves between facing gold electrode contacts.

since it is the most sensitive to mass changes. Figure 6.2 schematically illustrates the structure of a BAW device or crystal microbalance.

For AT-cut quartz crystals operating in the shear mode, the oscillation frequency f_o is inversely proportional to the thickness d of the crystal, as described by

$$f_o = \frac{k_f}{d} \tag{6.1}$$

where k_f is the frequency constant; for example, for an AT-cut quartz crystals at room temperature, k_f is 0.168 MHz/cm. Since the mass of the crystal m is $\rho A d$, where ρ and A are the density and cross-sectional areas, respectively, Eq. (6.1) indicates that the oscillation frequency is inversely proportional to the mass of the crystal.

This mass sensitivity can be exploited to fabricate chemical sensors by immobilizing a film on the resonator's surface that selectively binds a target analyte. When a species binds to the chemically sensitive film, it increases its mass by Δm, causing the resonance frequency to shift by Δf according to the Sauerbrey equation.

$$\Delta f = -\frac{1}{\rho_m k_f} f_o^2 \frac{\Delta m}{A} \tag{6.2}$$

where ρm is the density of the thin active coating on top of the piezoelectric substrate.

The quartz crystal microbalance is a very stable device that is capable of measuring extremely small mass changes. For example, for an AT-cut quartz plate with a resonance frequency of 5 MHz, a shift in resonance frequency of 1 Hz can easily be measured which corresponds to a change in mass of just 17 ng/cm^2.

As with many sensors, the performance of mass-sensitive devices is not instrumentally limited but depends strongly on the properties of the chemically sensitive film immobilized on the resonator. We consider the importance of the film properties later.

6.2.3. Surface Acoustic Wave (SAW) Devices

Unlike bulk acoustic wave devices, SAW detectors work by interaction of a wave traveling down the surface of the device with chemicals on or near the surface. An attractive phase-sensing device is the surface acoustic wave (SAW) oscillator employing interdigitated electrodes. Figure 6.3 shows that the acoustic wave is created by an AC voltage signal applied to a set of interdigitated electrodes at one end of the device. The electric field distorts the lattice of the piezoelectric material beneath the electrode, causing a surface acoustic wave to propagate toward the other end through a region of the crystal known as the *acoustic aperture*. When the wave arrives at the other end, a duplicate set of interdigitated electrodes generate an AC signal as the acoustic wave passes underneath them. The signal can be monitored in terms of amplitude, frequency or phase shift. These devices operate at ultrahigh frequencies (gigahertz range), giving them the capability to sense as little as 1 pg of material.

The interdigitated electrode behaves as a sequence of ultrasonic sources. For an applied sinusoidal voltage, all vibrations interact constructively provided that the distance between adjacent fingers is equal to half the elastic wavelength. The frequency of the voltage sine wave causing this effect is the synchronous or resonance frequency. Destructive interference between the elastic waves is observed at other frequencies, and the overall vibration is weaker, rendering the device less sensitive to mass changes.

6.2.4. Chemically Sensitive Interfaces

Having established the basic principles of piezoelectric sensor operation, we now focus our attention on the chemically sensitive film that transforms the nonselec-

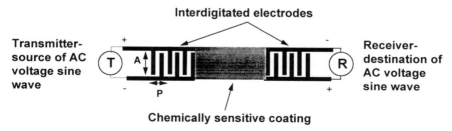

Figure 6.3. Layout of a single acoustic aperture surface acoustic wave (SAW) device.

tive device into a sensor that ideally responds only to a single analyte, but at worst responds only to a small group of analytes. To make these piezoelectric devices sensitive to specific species of chemical, biological, or environmental interest, they must first be modified with a layer with which only the species of interest will interact. Modification can be achieved using either synthetic or biological materials, but more chemically and physically robust sensors are formed by using synthetic materials such as polymers, inorganic complexes, and zeolites. However, biological materials, such as enzymes and antibodies, exploit the tremendous specificity of these materials to give sensors whose responses are highly specific for a single species even in complex media.

6.2.4.1. Film Deposition

To generate a change in mass, the target analyte must either adsorb onto a film deposited on the surface of the device, or the analyte must partition into the bulk of the coating. Adsorption interactions range from physical adsorption of the analyte (physisorption), to chemical bonding of the analyte to the modified interface (chemisorption). The weak interactions involved in physisorption generally produce only weak selectivity toward the target analyte. However, the sensor response is reversible, and there is no hysteresis when the device is exposed to a series of samples containing increasing, and then decreasing, concentrations of analyte. The stronger, although less reversible, interaction of chemisorption yields more highly selective sensors, but they may be usable for only a few measurement cycles.

Chemically sensitive films may be immobilized using a wide variety of techniques. Popular procedures involve simple droplet evaporation, spraying, or spin coating. These methods require a solution of the coating material to be prepared in a suitable volatile solvent. In the spray technique, the solution is aspirated to generate a finely dispersed aerosol that can be deposited onto the device at a controlled rate. Droplet evaporation is similar, except that a known volume of a particular concentration is directly applied to the crystal, and the solvent allowed to evaporate. Spin coating uses a more concentrated, and therefore viscous, solution that is either dropped onto the crystal surface while it is rotating, or is first applied, and then the device is rapidly spun. Spin coating can given films of micron thicknesses that are extremely uniform. Alternatively, more complicated modification techniques involve immobilizing Langmuir–Blodgett films, physical and chemical vapor deposition, or electrodeposition. The most stable films are obtained where the chemically sensitive film is chemically, rather than physically, bonded to the acoustic sensor.

It is important to note that in the case of SAW devices, the film should be deposited not only over the acoustic aperture but also over the interdigitated electrodes if the theoretical sensitivity is to be realized. However, this may not always be possible; for example, depositing an electronically conducting polymer or metallic film would short-circuit the transmitter and receiver electrodes.

Bioactive components, that will interact with the analyte of interest in such a way that the surface mass density is changed, can also be immobilized on the sensor surface. The most important bioactive surfaces for mass-sensitive devices employ an affinity binding-type mechanism using antibodies, lectins, receptor proteins, or nucleic acids. Antibody–antigen reactions are widely used as selective binding agents in immunoassays. Antibodies can be produced to selectively bind medium to high-molecular-weight materials including drugs, microrganisms, nucleic acids, hormones, and other proteins. However, it is difficult if not impossible to obtain antibodies that selectively bind small molecules. Other biomaterials also show promise for sensor development; for instance, lectins can be used to selectively bind carbohydrates, while immobilizing single strands of DNA on the surface of a piezoelectric device makes it possible to detect complementary strands of DNA. This detection scheme is especially applicable to the detection of viruses.

Biological materials can be confined to the surface of a piezoelectric device by either physical adsorption or chemical binding. For example, an IgG-type antibody will physically adsorb onto a silica surface that has been made hydrophobic by treatment with dimethyl dichloro silane, or onto a silica surface rendered hydrophilic using aminiopropyl trimethoxy silane. However, only the hydrophobic surface modified with IgG will selectively bind its corresponding antigen. This observation highlights a disadvantage of physical adsorption in that binding results in a random orientation of the antibody on the surface, which can cause blocking of the active site for binding. Furthermore, physical adsorption is often reversible, and sample solutions may wash the physisorbed material off the mass-sensitive surface.

Covalent bonding of bioactive material to the piezoelectric crystal avoids the difficulty of loosing the active film when it is rinsed between samples, and it may be possible to control the orientation of the immobilized molecules. Biological materials are covalently attached by chemical reactions between functional groups on the biomolecule and the sensor surface. There are many functional groups that can be used for surface attachment, including amines, aldehydes, sulphonyl chlorides, isocyanates, thiols, phenols, carboxylic acids, epoxides, and diazonium salts. Since the surfaces of piezoelectric crystals, such as quartz, contain large numbers of oxides, silane chemistry is often used to activate the crystal's surface so that the biomaterial can be reacted with it.

Beyond physical and chemical adsorption, a third approach to immobilizing bioactive materials is to encapsulate them within a polymer film at the sensor surface. A polymer film can be grown by first attaching a reactive monomer layer to the crystal's surface. The biomaterial can be immobilized by placing it in solution as the polymerization proceeds, or by allowing it to diffuse into a preformed film. The stability of either film can be improved by crosslinking so as to entrap the bioactive substance, and to improve the physical properties of the polymeric coating. Polymer films are much thicker than the monolayers formed by adsorption, which tends to increase the response time of the sensor.

While the coating thickness dictates the response time of the sensor, it is the density of active sties for analyte binding per unit area that is the critical factor controlling the sensitivity. Higher densities of active material will give larger mass changes for a given concentration of analyte, making the sensor more sensitive and giving the device a lower limit of detection.

6.2.4.2. *Chemical Sensing*

Many chemically sensitive acoustic wave devices are used for detection in the gas phase. For example, a representative response of a SAW device to water vapor illustrated in Figure 6.4. Exposing the device to water vapor causes it to partition into the immobilized polymer film increasing its mass and thus decreasing the resonance frequency. When the source of water vapor is removed, the water content of the film reequilibrates with its surroundings and the resonance frequency returns to its original value if there is a sufficiently long time interval between samples. Reversible responses of this type are essential for continuous monitoring.

The response time of these sensors is typically controlled by the interaction of the analyte with the coating rather than by the transducer itself. Even for polymer films with thicknesses of the order of microns, a fairly rapid response to a change in analyte concentration is observed. For example, in Figure 6.4 the response reaches 90% of its equilibrium value within 100 s. The response time depends on the analyte concentration with equilibrium responses being achieved more rapidly at higher analyte concentrations. Numerous properties of the immobilized films influence how rapidly these sensors respond to changes in the concentration of the target analyte. These properties include the film thickness, morphology, internal free volume, viscosity, and chemical structure.

Systems such as the one discussed above involving partitioning of an analyte into the bulk of a chemically sensitive coating, have been used to develop sensors for a wide variety of analytes. Table 6.3 shows how exposure of 19 ppm strych-

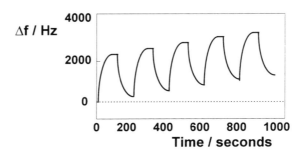

Figure 6.4.   Typical response of a SAW device coated with poly(ethyleneimine) operating at 158 MHz exposed to consecutive pulses of 1000 mg/m^3 water vapor at room temperature.

Table 6.3. Adsorption Masses and Partition Coefficients P of Strychnine and β-Ionene for Various Films Immobilized on a QCM[a]

Immobilized Film	Strychnine		β-ionene	
	Δm (ng)	P	Δm (ng)	P
Uncoated	2	10	2	10
$2C_{18}N^+2C_1$/PSS[b]	533	2700	610	3050
DMPE[c]	560	2800	540	2700
Poly(vinyl alcohol)	4	18	4	19
Poly(methyl glutamate)	5	25	6	30
Poly(styrene)	7	35	7	35
Bovine plasma albumin crosslinked with glutalaldehyde	5	25	6	30
Keratin	7	35	6	30

[a]The measurements were performed at 45°C.
[b]PSS = poly(styrene sulfonate).
[c]DMPE = dimylistoylphosphatidylethanolamine.

nine (a pesticide) or β-ionene (an odor) affects the adsorption masses of quartz crystal microbalances coated with various chemically sensitive films. This table also includes data on the partition coefficient P, specifically, the ratio of analyte concentrations inside and outside the film. A high partition coefficient indicates that the analyte partitions preferentially into the film immobilized on the microbalance.

The ability of these coated quartz crystal microbalances (QCMs) to selectively determine strychnine of β-ionene is revealed by examining the ratio of their respective partition coefficients. Inspection of Table 6.3 reveals that while coatings of $2C_{18}N^+2C_1$/PSS$^-$ or DMPE give large mass changes when exposed to either strychnine or β-ionene, the similarity of their partition coefficients effectively precludes the selective determination of one of these compounds in the presence of the other; specifically, a sensitive, yet poorly selective, response is obtained. In general, systems relying on analytes partitioning into a polymer film exhibit good selectivity between different classes of compounds, but have poor discriminating power between chemically similar materials such as individual members of a homologous series.

An example of chemisorption of an analyte onto a coating involves a film of square-planar platinum complexes of ethylene that has been used to measure the concentration of vinyl acetate. In this sensing scheme, vinyl acetate in the surroundings displaces the ethylene in the original complex, resulting in an increase in mass loading. This approach is essentially irreversible since displaced ethylene is lost to the surroundings. Therefore, the polymer film coating must be periodically regenerated by exposing the device to a high concentration of ethylene.

Temperature can play an important role in determining the behavior of piezo-electric devices. However, with advanced oscillator design and improved packaging, sensitivities of less than 1 Hz per degree centigrade can be easily achieved. Temperature not only affects the intrinsic transduction mechanism but will also influence the response time and absolute frequency change brought about by exposing the device to a particular analyte concentration. This sensitivity to environmental temperature is a particular problem for piezoelectric crystals coated with thick films into which the analyte must partition. The strong temperature dependence arises because the rate of diffusion of vapor into the coating increases with increasing temperature. Moreover, the equilibrium concentration of vapor within the coating decreases with increasing temperature, and the corresponding frequency change is therefore reduced at higher temperatures.

Irrespective of the synthetic approach used to immobilize a chemically sensitive film on these piezoelectric crystals, or the mechanism by which the analyte interacts with the coating, one must ensure that the film is sufficiently thin and rigid to avoid viscoelastic effects when they are used for sensing in solution. For example, it has been found that responses to glucose binding to the hexokinase enzyme entrapped in polyacrylamide films are far larger than expected simply on a mass-change basis. These discrepancies almost invariably arise from changes in the film structure accompanying analyte interaction with the film. Despite these viscoelastic effects, the frequency change is often proportional to the analyte concentration making them potentially of analytical value.

Given the small size and ease with which these piezoelectric devices can be manufactured using planar silicon processing techniques, it is practical to combine several of these devices to form a sensor array. By coating each element of the array with a different material selected for favored partitioning of a particular component of the sample matrix, one obtains a slightly different response pattern from the array even for mixtures containing compounds with similar structure. These data can then be processed using pattern recognition or chemometric approaches to deconvolve the responses due to interference and target analytes. Sensor arrays of this type are considered in greater detail later.

6.2.5. Conclusions

Mass-sensitive devices will continue to be used as the basic transduction mechanism of a wide variety of chemical sensors because of their tremendous sensitivity. Their sensitivity will continue to improve by using sensors that operate at increasingly high frequencies and by new device structures. However, new chemically sensitive coatings promise the greatest improvement in overall sensor performance. For example, in gas sensing the best organic polymer coatings trap only about 1% of the vapor molecules presented to them. Therefore, there are substantial opportunities for enhanced performance in terms of the coating–analyte interaction.

It is also likely that the ability of mass-sensitive devices to provide real-time information about an analyte's concentration will be increasingly exploited. This ability will be important not only for analytical applications but also for studying the rates and mechanisms of solid–gas and solid-liquid reactions, and for characterizing the properties of materials such as the diffusion rates of small molecules through polymers.

6.3. MICROAMPEROMETRIC SENSORS

6.3.1. Introduction

In the following section, we move away from mass-sensitive devices and consider sensors that detect the presence and concentration of electroactive species based on reduction–oxidation, or redox, reactions. Microelectrodes, also commonly known as *ultramicroelectrodes,* may be defined as electrodes whose critical dimension is in the micrometer range, although electrodes with radii as small as 10 Å have been fabricated. These small amperometric sensors have greatly extended the range of sample environments and experimental timescales that are useful for electroanalysis. In this section, we explore some of the exciting and innovative practical applications of these electrodes whose active surface areas are many times smaller than the cross section of a human hair.

Microelectrodes have several desirable attributes, including small currents, steady-state responses, and short response times. The currents observed at these sensors typically lie in the picoampere–nanoampere (pA–nA) range, which is several orders of magnitude smaller than those observed at the conventional macro-electrodes considered in detail by Cassidy, Doherty, and Vos in Chapter 3 (of this volume), where the radius is usually several millimeters. These reduced currents are a key element in the successful application of microamperometric sensors. In the past, the range of conditions under which electrochemical measurements could be made was restricted to highly conducting media, such as aqueous electrolyte solutions. This restriction arose because resistance between the working or sensing electrode, and the reference electrode, limited the precision with which the applied potential could be accurately controlled. The small electrolysis currents observed at microamperometric sensors often completely eliminate these ohmic effects. The immunity of microelectrodes to ohmic drop phenomena allows one to quantify the concentrations of electroactive analytes in previously inaccessible samples such as nonpolar solvents, supercritical fluids, and even solids.

The small size of these sensors makes diffusional mass transport extremely efficient. In fact, mass-transport rates to a microelectrode are comparable to those of a conventional macroelectrode that is being rotated at several thousand revolutions per minute. As discussed in Chapter 3, at relatively long experimental timescales, the dimensions of the diffusion layer exceed the radius of the microelectrode, and the originally planar diffusion field transforms into a spherical diffusion field..Consequently, the flux of electroactive species to the electrode is substantially higher than for the pure planar diffusion case that is typical of a macroelectrode. This efficient mass transport allows one to observe steady-state responses when the applied potential is slowly scanned in cyclic voltammetry. The sigmoidal shaped responses observed in these experiments are analogous to the polarograms obtained using a dropping-mercury electrode, or a rotating-disk electrode, but they are observed under entirely quiescent conditions. The steady-state limiting current is directly proportional to the analyte concentration, making it extremely useful for determining the concentration of analytes in liquid, solid, and even gas phases.

6.3.2. Microelectrode Geometries

Before discussing how microelectrodes are used as chemical sensors, we first consider how they are constructed. Figure 6.5 illustrates the five common microelectrode geometries. The microdisk is the most popular geometry, and is employed in approximately 50% of all investigations. Other common geometries include cylinders (20%), arrays (20%), with the remaining 10% comprising bands, rings, and less frequently spheres, hemispheres, and more unusual assemblies. The most popular materials include platinum, carbon fibers, and gold, although mercury, iridium, nickel, silver, and superconducting ceramics have also been used. Microdisk electrodes predominate because of their ease of construction, and the ability to mechanically polish the sensing surface. Microelectrodes in the form of disks, cylinders, and bands are commonly fabricated by sealing a fine wire or foil into a nonconducting electrode body such as glass. Microlithographic techniques are perhaps the best method of producing well-defined microelectrode arrays. Other array fabrication methods include immobilization of large numbers of metal wires within a nonconducting support, and electrodeposition of mercury and platinum within the pores of a polymer membrane. Spherical and hemispherical microelectrodes are typically formed by electrodepositing mercury onto platinum or iridium microdisks.

6.3.3. Properties of Microelectrodes

6.3.3.1. Reduced Capacitance

When an electrode comes into contact with an electrolytic solution, a double layer is formed at the interface in which the charge present on the metal electrode is

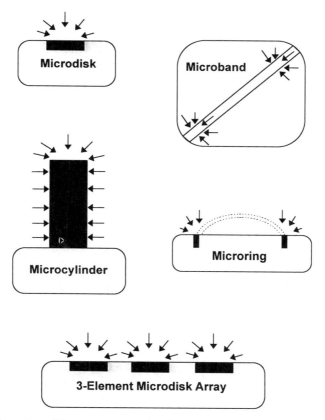

Figure 6.5. Illustrations of the most common microelectrode geometries and their diffusion fields.

compensated for by a layer of oppositely charged ions in solution. In many respects, this electrochemical double layer behaves like an electrolytic capacitor. In particular, when the applied potential is changed, a current flows to charge the double-layer capacitance. However, when using a microamperometric sensor, it is the magnitude of the Faradaic current, corresponding to oxidation or reduction of the analyte, that is used to determine its concentration. The presence of a charging current complicates this measurement since the Faradaic current must be measured on top of this background current. Therefore, for quantitative measurements, one seeks to minimize this charging current. The electrode capacitance is an extensive property, and is proportional to the electrode area. Thus, by shrinking the size of amperometric sensors from millimeters to micrometers, one can observe greatly reduced total capacitances.

6.3.3.2. Ohmic Effects

When Faradaic and charging currents flow through a solution, they generate a potential that acts to weaken the applied potential by an amount iR, where i is the total current and R is the cell resistance. This can lead to severe distortions of the experimental response. Microsensors may significantly reduce these ohmic effects because the currents observed are typically six orders of magnitude smaller than those at macroelectrodes. These small currents often completely eliminate iR problems, even when performing an analysis in a highly resistive organic solution such as chloroform or even hexane.

6.3.4. Electroanalysis

While microelectrodes can help identify compounds by accurately measuring their formal potentials, they are used predominantly to determine analyte concentrations. In voltammetric experiments, when the applied potential is changed slowly in time (<50 mV/s), the resulting current due to reduction or oxidation of an electroactive species exhibits a sigmoidal-shape response. The steady-state limiting current observed is directly proportional to the concentration of the electroactive species. Therefore, if one knows the diffusion coefficient, it may be possible to determine the analyte concentration without the need for a calibration curve.

The low currents, high sensitivity, and relative immunity of microamperometric sensors to ohmic effects greatly simplifies electroanalysis. These attributes mean not only that simpler instrumentation can be used (e.g., two-electrode instead of three-electrode potentiostats) but also that microelectrodes can be used for electroanalysis in media of high electrical resistivity, such as soil, foodstuffs, and solutions without deliberately adding a supporting electrolyte.

Thin mercury films deposited on ultrasmall carbon-ring electrodes have been used in anodic stripping voltammetry (ASV) experiments to determine Pb^{2+} concentrations in solution. Thin ring microelectrodes can have effective diffusional areas that are more than 100 times larger than microdisks of the same geometric area. This increase in accessibility to a diffusing species gives a higher current efficiency that can reduce the limit of detection by an order of magnitude. Performing ASV in the absence of deliberately adding supporting electrolyte was investigated as means of reducing impurity levels in the samples. It is possible to perform meaningful electrochemistry using microelectrodes because, as discussed above, their responses are relatively impervious to ohmic effects. Figure 6.6a shows the ASV response for a solution containing 20 μM Pb^{2+} that was recorded at a thin mercury film, where the supporting electrolyte was 0.1 M KNO_3. The lead amalgam is oxidized at -0.46 V, followed by an unknown impurity at 0.0 V, and mercury at $+0.41$ V. Figure 6.6b shows that in the absence of Pb^{2+}, or deliberately added supporting electrolyte, only stripping peaks due to oxidation of an impuri-

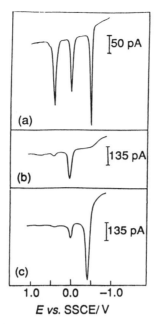

Figure 6.6. Anodic stripping voltammograms at a thin mercury film deposited in situ on a 2-μm carbon-ring electrode for (a) 2.0×10^{-7} M Pb^{2+}, 1.0×10^{-6} M Hg^{+}, and 0.1 M KNO_3 (pH 3.0) as supporting electrolyte; (b) only 1.0×10^{-6} M Hg^{+} (blank); and (c) 2.0×10^{-7} M Pb^{2+} and 1.0×10^{-6} M Hg^{+} (i.e., no supporting electrolyte). Preconcentration potential was set at -0.9 V for 300 s. Scan rate was 400 mV/s. [Reproduced from D. K. Y. Wong and A. G. Ewing, *Anal. Chem.* **62,** 2697 (1990), with the permission of the American Chemical Society.]

ty at 0.0 V, and mercury at $+0.40$ V, are observed. Figure 6.6c shows that in the absence of KNO_3, the signal due to the impurity is reduced relative to the Pb^{2+} response. Furthermore, an enhanced sensitivity, that is, a larger absolute peak current, is observed for a given Pb^{2+} concentration when electrolyte is not deliberately added.

The concentration of Hg^{+} used for in situ deposition of the mercury films also affects the analytical performance. For low concentrations of Hg^{+} (μM), the stripping current for Pb^{2+} does not decrease with decreasing lead concentration as rapidly as expected, thus giving enhanced sensitivity and lower detection limits. It is possible to determine Pb^{2+} at concentrations as low as 3.2×10^{-11} M using this method.

The high analyte flux at microelectrodes, and their short response times, can be exploited to increase the speed and sensitivity of stripping analysis. For example, fast cathodic stripping analysis at microelectrodes has been used to determine various anions including iodide, bromide, sulfide, and cysteine. The results obtained using electrodes of conventional size at slow scan rates ($\sim$100 mV/s) were compared with those obtained using microelectrodes under fast linear scan conditions ($\sim$700,000 mV/s). Hemispherical microelectrodes of radii between 2.5 and 12.5 μm were fabricated from silver or amalgamated copper, gold, or platinum. Studies of this type suggest that fast scan methods employing microamperometric sensors can simultaneously decrease the analysis time, and improve sensitivity since larger currents are observed at high scan rate.

The immunity of the microelectrode voltammetric response to convection, and their small physical size, have been exploited to give sensitive detectors in high-performance liquid chromatography (HPLC) and capillary electrophoresis. Direct amperometric detection within small-bore capillaries offers a sensitive detection method that avoids difficulties inherent in other systems. For example, on-column UV detectors cannot be used with narrow capillaries because, due to a much shorter path length, the detection sensitivity is drastically reduced. Capillary electrophoresis with electrochemical detection in 2- and 5-μm capillaries has been developed to study ultrasmall biological environments. Sample volumes as small as 270 fL have been directly injected from the cytoplasm of a single nerve cell of the pond snail *Planorbis corneus*. It is possible to obtain subattomole detection limits for easily oxidized species such as serotonin, using an etched carbon fiber microelectrode located within the capillary.

6.3.4.1. *Biological Systems*

The critical dimension of a microelectrode is typically in the 0.1–50-μm range. However, many fabrication methods give electrodes in which the sensing area is microscopic, but the complete electrode is macroscopic because the nonconducting body has a radius of several millimeters. These electrodes are not useful for performing electrochemistry in small volumes, or for obtaining information about redox activity with high spatial resolution. Therefore, other encapsulation methods have been developed to ensure that the nonconducting material is thin. One method involves insertion of carbon fibers into microscopic tapered glass pipettes that are subsequently sealed with epoxy resin. An active electrode surface is subsequently exposed by mechanical polishing, to give an elliptical microelectrode in which the minor axis is approximately 5 μm and the major axis is approximately 35 μm. An alternative procedure involves electropolymerization of a passivating polymer film around the carbon fiber electrode. Both of these methods can give electrodes with total diameters in the tens-of-micrometers range.

These small electrodes are widely applied in studies of biological systems since their implantation causes little tissue damage, yet they still provide a sufficiently large area for sensitive extracellular measurements. The microprobes offer a relatively noninvasive means of in vivo monitoring, not only because they are physically small but also because of the minute quantities of material electrolyzed. In vivo monitoring of biologically important species such as catecholamines (dopamine, norepinephrine, and epinephrine), ascorbic acid, 5-hydroxytrypta-mine, uric acid, and oxygen has been achieved using these amperometric microsensors. Furthermore, pharmakinetic information, including metabolic pathways, assimilation rates, and therapeutic levels, has been obtained for common drugs such as aspirin and theophyline.

In vivo monitoring of dopamine with the mammalian brain can provide direct, real-time, quantitative information about brain chemistry, Parkinson's disease, and the action of antipsychotic drugs. Slow-scan-rate voltammetry at carbon fiber microelectrodes has been widely used to quantify dopamine concentrations in the grains of anaesthetized rats. Problems with interferences from other species, such as ascorbate and dihydroxyphenylacetic acid (DOPAC), that are redox active in the same potential region, can be attenuated by modifying the microelectrode surface with a cation-exchange membrane such as Nafion. This coating serves the dual purpose of improving selectivity and reducing electrode fouling. These studies demonstrate that, while the dopamine concentration within brain tissues is in the micromolar range, its concentration in extracellular fluid is extremely low. This finding has subsequently been confirmed by dialysis studies that indicate an extracellular dopamine concentration of less than 10 nM. A fast-scan-rate voltammetric method has been used to provide temporal information about the firing rate of dopamine neurons in response to external chemical stimulation. The dopamine concentration can be determined on a 100-ms timescale by scanning the applied potential at 300 V/s. These measurements demonstrate the transient nature of chemical changes occurring during brain activity. Furthermore, one can probe the characteristics of chemical transmission between neurons in different brain regions by determining apparent rate constants.

6.3.5. Conclusions

Microelectrodes with radii in the micrometer range, but whose surface is modified with spontaneously adsorbed, or self-assembled redox active monolayers, will form the basis of new families of sensors. It appears likely that new electroanalytical techniques will emerge that exploit the high-quality kinetic information that microelectrodes provide at short times. At present, selectivity in the electroanalysis of multicomponent systems is achieved on the basis of individual redox active species displaying different formal potentials, or interacting selectively with a modifying film. Furthermore, successful analysis typically relies on measuring the limiting steady-state current in slow-scan-rate cyclic voltammetry that is proportional to the analyte's concentration. Therefore, in common with many analytical techniques, achieving a selective response depends on thermodynamic differences between species (i.e., different formal potentials), and achieving a steady-state condition. However, in the future, microelectrodes may be applied to determine analyte concentrations based on differences in their reactivity rather than differences in their formal potentials.

6.4. SENSOR ARRAYS AND SMART SENSORS

6.4.1. Introduction

In common with all analytical tools, sensors have problems with their implementation. Two particular problems are selectivity and ruggedness. While the greatest area of interest in sensors has been to increase the number of target analytes for which chemically sensitive films and coatings exist, there has also been a drive toward sensors having enhanced functionality. This has occurred predominantly by integrating sensors into arrays to give a device capable of measuring the concentration of several different analytes or to provide more robust analytical systems. The advantages of arrays over discrete sensors include the ability to use partially selective sensors and to analyze multicomponent samples. This approach gives the capability for direct sensing of a specific analyte in the presence of an interfering background or simultaneous monitoring of two or more analytes. Beyond these considerations, considerable attention is being paid to developing "smart" sensors with greatly improved abilities to process analytical information. A smart sensor exhibits the following characteristics:

- Ability to process information for automatic calibration or compensation of baseline drift.
- Ability to communicate with other sensors or devices.
- Adaptability to changes in its environment, such as automatic temperature compensation in optical or amperometric measurements.
- Ability to self-diagnose performance.

A self-diagnosing intelligent sensor would be able to provide the operator with information about its own performance and, if appropriate, modify its own internal calibration data to maintain the accuracy of the measurements. Alternatively, smart sensors can inform the operator of abnormal operating conditions, ensuring that all analyses are performed uniformly; warn of impending failure; and, if necessary, remove themselves from the measurement cycle. The ability of a sensor to easily exchange information with another device is especially important in the shift toward integrated manufacturing facilities in which there is a free flow of information around the factory. All of these functions demand that the sensors have decision making capabilities. In the past, this has been achieved using parametric data; for example, a sensor interfaced to a microprocessor continuously monitors the pH of a fermentation vat and warns an operator if the pH exceeds some previously specified limit stored in the processor. While this approach will continue to be used, there is a trend toward using nonparametric methods such as neural networks, fuzzy logic, and expert systems to evaluate the significance of sensor signals.

6.4.2. Integrated Signal Processing

Directly integrating a sensor element and a signal processing unit within a single chip promises more sensitive detection since integration frequently offers much greater immunity to electromagnetic interference, thus improving the signal-to-noise ratio. For example, the potentiometric ion-selective electrodes considered in Chapter 2 are high-impedance sensors, and the leads carrying their responses are prone to corruption by stray capacitance and electrical noise. These effects mean that the sensor outputs cannot be transmitted over long distances. However, by integrating the sensor with a circuit that converts the high-impedance sensor response into a low-impedance electrical signal, the stability, sensitivity, and resolution of the sensor can be dramatically improved.

6.4.3. Self-Testing and Self-Calibrating Sensors

Sensors capable of diagnosing their own performance are highly desirable. Capabilities in this area range from simply confirming that the sensor is functioning correctly, to systems that use the current measurements to dynamically update their calibration information. Complete failure of the device is usually detected by a sensor output, typically either current or voltage, that falls outside the acceptable range. However, a more sophisticated diagnostic capability is preferable since many sensors give a reasonable output yet fail to perform adequately in terms of sensitivity or selectivity. For example, for amperometric microsensors it ought to be possible to diagnose fouling of the electrode surface by measuring the electrode capacitance since this will typically decrease when the surface of the sensor becomes fouled. It may also be possible to recognize sensor malfunction by periodically probing a noise characteristic, such as spectral power density in the case of optical sensors.

While self-diagnosis leading to go/no-go decisionmaking, is a significant advance over "blind" sensors, it is preferable to have sensors that can update their calibration information on the basis of the current operating conditions. For example, the responses of many of the sensors discussed in this book, including the potentiometric (Chapter 2), amperometric, and optical sensors (Chapter 5), are sensitive to the ambient temperature. This effect can be removed by integrating the chemical sensor with a device, such as a thermocouple or thermistor, that independently measures the environmental temperature. By using an appropriate compensation algorithm, one can then use these temperature readings to correct the outputs of the chemical sensor.

6.4.4. Sensor Arrays

As illustrated schematically in Figure 6.7, arrays of chemical sensors have been fabricated with the objective of improving the quantity, quality, and reliability of the analytical information produced. Each individual sensor, perhaps integrated with its own signal processing or conditioning circuit, is connected to an intelligent processor that takes the individual outputs and analyses the response pattern to extract analytically useful information. On a basic level, the array may consist of several identical sensors, and the artificial-intelligence unit may simply poll the individual responses so as to identify malfunctioning sensors. For example, an array of 10 identical pH probes has been developed, and with identification of probes whose output does not match the majority response, self-diagnosis of the sensor array is accomplished. Apart from the advantage of self-diagnosis, individual probes may provide information about the chemical composition at physically different locations within the sample. Applications of this distributed, rather than integrated, sensor array technology include emission monitoring around factories.

   While the redundancy offered by arrays of identical chemical sensors is attractive for improved self-diagnosis and overall reliability, arrays containing probes

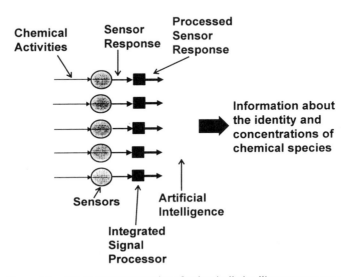

Figure 6.7.   Schematic representation of a chemically intelligent sensor array.

that respond to different chemical species promise to revolutionize chemical sensing. A first-order sensor array is a collection of n partially selective sensors that are operated together when analyzing a sample. The objective is to make the $1 \times n$ response for a particular analyte unique. Therefore, each sensor must have some discriminating power between analyte and interferences, but the outputs from each individual sensor must different from one another. Examples of first-order sensor arrays include quartz crystal microbalances (QCMs), metal oxide semiconductor (MOS) sensors, field-effect transistors (FETs), and amperometric sensors. Improved accuracy and precision can be achieved by increasing the discriminating power of the array, namely by choosing sensing elements that are more perfectly selective toward one analyte.

Potentiometric ion-selective electrodes also make excellent sensors for constructing first-order arrays. The response of these chemical sensors is typically selective, in that they generate a signal when exposed to a particular family of chemical species, such as alkali-metal cations, rather than responding specifically to a single chemical species, such as sodium cations. A four-electrode array comprising three highly selective and one sparingly selective potentiometric sensor offers improved performance for the determination of sodium, potassium, and calcium ions in tertiary mixtures. The array response, operated as a detector in flow-injection analysis, was modeled by the Nicolskii–Eisenman equation (Chapter 2) using experimental data obtained from calibration solutions selected according to a factorial experimental design. Nonlinear optimization methods were used to evaluate the standard cell potential, electrode slope, and selectivity coefficients for both of the interfering ions. In those circumstances where there is a well-developed theoretical model of the sensor response, this deterministic approach is preferable to artificial intelligence approaches based on neural networks of fuzzy logic since malfunctioning sensors can be easily identified.

Including a sparingly selective sensor within the array has both advantages and disadvantages. In mixtures containing only sodium, potassium, and calcium ions, that is, no unmodeled interferents, the advantages over traditional single-electrode measurements include polling of predictions, thus improving the accuracy and precision of the analysis, and the array can diagnose its own performance without recalibration. Unlike traditional single-electrode approaches, the array could accurately determine low levels of individual cations in the presence of a large and widely varying excess of the other two. This improved performance is attributable to the accuracy with which the cell potentials, the selectivity coefficients, and electrode slopes can be determined using the nonlinear modeling procedure compared to that achievable using traditional methods. However, in solutions that contain ions not included within the model, arrays including the sparingly selective sensor do not perform as well as an array containing highly selective sensors alone. This deterioration in response arises because the model cannot decouple those signals arising from the analyte and unmodeled interferents. This result highlights a difficulty in using sparingly selective sensors in arrays. In complex mixtures, at least one additional term, which may not add simply to the model, will have to be included for each likely interference if the array response is to be accurately modeled. Failure to include likely interferents will result in inaccurate determinations where the relative concentrations of interferents is variable. In addition to this, complete characterization of the response surface of an array of sparingly selective sensors will require a large number of calibration solutions.

Including a nonspecific electrode within a potentiometric sensor array has been further developed by immobilizing several highly selective ionophores within a single sensor so that it responds to a range of analytes, yet the number of interferents not included in the model of the array response is minimized. This approach exploits the advantages of chemical selectivity and the ability of the nonlinear model (system of Nickolskii–Eisenman equations) to decouple the analyte and interferent signals arising from the selective rather than specific sensor responses.

This work on sensor arrays interfaced to a computer also showed that the sensor responded relatively more rapidly to its primary ion than to an interferent. This result suggests that data obtained under kinetic rather than equilibrium conditions may offer an enhanced analytical performance. However, for this advantage to be realized, very reproducible sampling, flow dynamics, and the fabrication of sensors with identical response characteristics will be required.

First-order sensor arrays have also been used in attempts to mimic the vertebrate olfactory system. Recent research suggests that the extraordinary sensitivity and selectivity of the sense of smell is achieved, not on the basis of highly specific binding sites, but from an extended array of sparingly selective binding sites within the nasal cavity. A major difficulty in trying to develop an *electronic nose* is that the response of gas and odor sensors are typically very poorly selective and respond to a wide range of compounds. However, various modeling approaches have been developed that give arrays of these sparingly selective sensors the following desirable properties that mimic natural olfactory systems:

1. A large sensor array.
2. The number of sensing elements stimulated depends on the odor concentration.
3. An invariant, stationary pattern is generated for each individual odor.
4. The outputs from individual sensing elements are independent from one another.
5. General (e.g., classification of odor as acrid, sweet or sour) and specific (identification of specific compounds present) information on the odor can be extracted.

The sensors used to form an array that generates a "signature" of each analyte range from surface acoustic wave, electrochemical, piezoelectric, and optical. There is a common theme to these sensing strategies in that each element of the array is functionalized, often with a polymeric coating, so that it exhibits both selectivity for many disparate chemicals (broadband response) yet provides high sensitivity. For example, by immobilizing dye molecules onto the tips of a fiber-optic bundle, a multianalyte optical sensor has been created that gives different fluorescent response patterns (spectral shifts, intensity changes, spectral shape changes, and temporal responses), which depend on the physical and chemical nature (polarity, shape, size, and functionality) of the organic vapor to which the array is exposed. This sensing strategy is distinguished from other systems where only a single parameter (signal intensity) is measured, such as differences in resonance frequency in acoustic-wave-based sensors. Multidimensional sensing approaches of this kind that measure many different parameters simultaneously ought to make it possible to produce a more sensitive, multianalyte sensor array with fewer elements. Table 6.4 summaries some of the types of gas sensor that have been applied specifically to the analysis of odors that are important in the food-and-drink industry.

While first-order arrays offer enhanced performance over single sensors, extending the array to a second dimension offers several unique advantages. For example, second-order arrays of $n \times p$ sensors can provide temporal information and are capable of quantitative information in the presence of unknown interferences. Examples of true second-order analytical systems include excitation–emission fluorescence and gas chromatography: mass spectrometry. These approaches are sec-

Table 6.4. Types of Chemical Sensor Arrays Used for Odor Analysis

Sensor	Number of Sensors within Array	Mode of Detection	Application
Conducting polymers	12	Conductiometric	Flavors in lagers and beers
Sintered metal oxide	12	Conductiometric	Aroma of coffee blends and roasts
Amperometric sensors	16	Amperometric	Toxic vapors in cereals
Metal field-effect transistors	20	Photocapacitive current	Odors of alcohols, ammonia, etc.
Lipid coatings	6	Mass-sensitive	Discrimination between alcoholic drinks

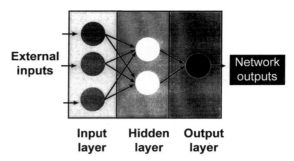

Figure 6.8.   Schematic representation of a neural network.

ond-order because each method is capable of independent measurement with some limited resolution of mixtures of components. It is remarkable that relatively few second-order sensor arrays have been developed. However, a fiber-optic sensor array that measures visible spectra (first dimension) and their time evolution (second dimension) has been developed. Separating analytical information in both the spectral and time domains offers much greater discriminating power than does either dimension alone.

Beyond the important issues of choosing the appropriate type of sensor and correctly designing the sensor array, considerable attention must be paid to the way that the complex response pattern of the array is analyzed. Parametric methods, such as linear and nonlinear regression or discriminant function analysis, represent one approach to analyzing array outputs. Mathematical regression allows a calibration model to be constructed that can then be used to resolve signals arising from different analytes in samples of unknown composition. Regression techniques range from linear models based on partial least-squares regression (PLSR) and principal-components regression (PCR) to nonlinear, nonparametric models based on multivariate adaptive regression splines (MARS) and projection pursuit regression (PPR). Pattern recognition approaches are also important, and these fall into two major categories: unsupervised (classification) and supervised (discrimination). These methods differ since a response model must first be established for supervised learning using a training set. Unsupervised learning, such as principal-components analysis (PCA), is frequently used for preliminary investigations designed to identify clustering of data in two-dimensional space.

However, for arrays containing large numbers of sensing elements or samples containing a large number of components, it is more usual to use nonparametric analysis, such as neural networks, pattern recognition, cluster analysis or back-propagation algorithms. Neural networks are meant to simulate brain function and are used to construct nonparametric, nonlinear models of the array response. Figure 6.8 is a schematic representation of a feedforward neural network. The network consists of three layers: an input layer, in which each unit is analogous to a predictor variable in regression; one or more hidden layers; and an output layer, in which each unit is analogous to the dependent variables in regression. All units have at least one input and one output. The output may simply be the sum of the inputs, but more usually a transfer function is applied to the sum, giving the neural network nonlinear modeling capabilities. In the hidden layer sigmoid functions are often used while linear functions are typical in the input and output layers. After multiplying by a weighting factor, the output of a unit is fed forward to a unit in the next layer.

The weighting factors are all randomized before training of the network commences. In the training procedure, calibration samples of known composition are presented to the sensor array and the responses entered into the network. The model is built by changing the weights within the neural network so that the differences between output units and the target values are minimized. In a practical application, this might mean that the sensor array/neural network combination correctly identifies the composition of all training samples when presented in a blind trial. Usually the training set must be presented several times before the weights no longer change when the complete training set is presented to the neural network. Although a considerable number of training algorithms have been developed, notably backpropagation and resilient propagation, this cumbersome optimization procedure is one of the most significant drawbacks of the neural networks. It is also very difficult to diagnose the way in which the trained neural network operates, for example, it is almost impossible to relate the weightings between layers and theoretical models to determine how the individual sensors should perform. That the sensors are essentially allowed to dictate their own model rather than attempting to fit the experimental data to a theoretical model also makes it more difficult to identify malfunctioning sensors; for instance, one cannot simply look at the slopes of individual ion-selective electrodes and identify failed sensors. Moreover, while neural networks generally perform well at qualitative tasks, such as identifying the presence of a particular analyte in a sample, they have been rather less successful at providing quantitative information, specifically, concentrations of specific analytes in complex samples.

6.4.5. Conclusions

There is a vast range of potential applications for sensor arrays capable of determining the components present within a particular gas or odor, as well as their concentrations. These range from the food-and-drink industry, pharmaceuticals and cosmetics, to environmental monitoring and the diagnosis of medical conditions. It has only recently become possible to manufacture large numbers of chemical sensors on a cost-effective basis. Therefore, it is now practical to construct sensor arrays that are sufficiently large to meaningfully improve the selectivity and sensitivity of chemical sensors. However, successful sensor arrays demand the parallel development of intelligent signal processing algorithms capable of modeling the outputs of sensor arrays.

6.5. TRENDS AND FUTURE DIRECTION

One of the major challenges to modern analytical chemists is the monitoring of environment and industrial processes. Modern chemical sensors make it feasible to continuously monitor a wide range of important analytes. While a vast number of discrete sensors and sensor arrays have been investigated, there is a definite need for an improved understanding of the underlying mechanisms that dictate sensor responses. Current developments involve the use of sensing materials that are better defined with respect to their physical structure and chemical composition. This advance is being achieved by using the full range of analytical techniques such as surface spectroscopy, atomic probe techniques, and X-ray crystallography, to characterize existing and potential materials for sensor applications. This systematic approach will yield sensors that are more reproducible, sensitive, and specific and have larger dynamic ranges, shorter response times, and longer lifetimes. Furthermore, this enhanced understanding ought to make it possible to identify new combinations of sensing materials that offer improved performance. These advanced sensors will be applied increasingly in environmental monitoring and medical diagnostics. Both of these areas demand sophisticated sensors that exhibit long-term stability, reliability, and ease of use, as well as reproducible and easily interpreted signals.

From a fabrication perspective, the major research drive is in the area of fabricating sensors using planar technologies. Therefore, there is intense interest in developing thin and thick films suitable for chemical sensing. Self-assembly, spontaneous adsorption, Langmuir–Blodgett Films, and chemical vapor deposition are all being used to produce films whose structure and chemical composition is carefully controlled and optimized for the detection of specific analytes. Applications of these materials include mass sensing, optical detection, and the development of selective membranes and barriers for potentiometric and amperometric sensing in liquids.

As discussed in the section on microamperometric sensors (Section 6.3), miniaturization will be a recurring theme. In particular, amperometric sensing in solids, nonpolar solvents, supercritical fluids, and gases will greatly diversify the range of media in which electroanalysis can be successfully performed. Beyond these possibilities, important future applications are likely to include studies of microamperometric sensors with sizes approximating molecular dimensions. With such advances, it may be possible to detect single molecules and to study adsorption and reorientation of single molecules.

SUGGESTED FURTHER READING

Forster, R. J., *Chem. Soc. Rev.,* **1994,** 289.

Vaihinger, S., and Goepel, W., in *Sensors, a Comprehensive Survey,* Goepel, W., Hesse, J., and Zemel, J. N., eds., VCH, Weinheim, German, 1991.

Yamamuchi, S., ed., *Chemical Sensor Technology,* Vol. 4, Kodansha Ltd., Tokyo, Japan, 1991.

CHAPTER

7

SENSOR SIGNAL PROCESSING

HUGH McCABE

School of Electronic Engineering, Dublin City University, Dublin, Ireland

7.1. INTRODUCTION

7.1.1. Discrete-Time and Continuous-Time Processes

A quantity being observed or measured by a sensor is given the generic title *signal*. For example, temperature measured over an interval of time at a given point constitutes a *signal*. Hence, if temperature is continuously monitored on a strip-chart recorder, then the continuous history of temperature versus time constitutes a continuous-time signal. On the other hand, if the temperature is measured at fixed time intervals, every hour, for example, then the series of discrete measurements constitutes a discrete-time signal.

Many processes from which signals are generated are continuous-time processes. However, since these signals are eventually stored and processed on digital computers, they must first be converted into equivalent discrete-time signals. Therefore, this chapter devotes an extensive discussion to discrete-time signals, both how they are represented and how they are processed. New concepts are first introduced in the continuous-time domain, however, because many sensor signals originate in this domain and also because these concepts are more easily grasped initially in the continuous-time domain. A continuous-time signal is referred to as an *analog signal,* and a discrete-time signal is referred to as a *digital signal.*

7.1.2. Further Reading

A list of references is provided at the end of this chapter. Collectively they discuss in greater detail the material presented here. They also present the material from a variety of different perspectives. Consequently, the reader is referred to these references as a whole rather than to any individual one.

7.1.3. Representing a Physical Process by a Signal

In general, a signal is represented as a function of time by some functional notation such as $x(t)$. For most signals encountered in practice, the time parameter t is usually greater than zero with $t = 0$ arbitrarily assigned to the beginning of the signal. For example, if we inject a chemical into a solution at some initial time $t = 0$, and if the concentration of the chemical decays exponentially from an initial value C_0 to some final value C, then the concentration $c(t)$ can be represented by the mathematical function

$$c(t) = C + (C_0 - C)\exp(-t), \qquad t > 0 \qquad (7.1)$$

If instead of injecting a chemical into a solution, we apply a battery of voltage V to an electrical RC circuit where the capacitor has an initial value V_0, it can be shown that the voltage $v(t)$ across the capacitor is given by

$$v(t) = V + (V_0 - V)\exp\frac{-t}{RC}, \qquad t > 0 \qquad (7.2)$$

Looking at Eq. (7.1) and (7.2), we observe that while they represent quite different processes, their mathematical representation in the form of signals $c(t)$ and $v(t)$ are remarkably similar. This, of course, is one of the powerful reasons for representing processes as signals. This mathematical abstraction from the process to the signal representation permits one to not only subsume whole classes of processes into a common mathematical framework but also to more fully identify and understand the basic parameters and forces at work in these processes. For example, in Eq. (7.2) we observe that the speed with which $v(t)$ converges to its steady-state value V is governed by the numerical value of the product RC. The smaller the value of RC, the more quickly $v(t)$ reaches its steady-state value V. Therefore, by a judicious selection of values for the resistance R and capacitor C, any desired response time can be achieved.

The processes expressed in Eqs. (7.1) and (7.2) are relatively simple ones. More complicated processes have correspondingly more complicated signal representations; however, notwithstanding the degree of mathematical complexity involved, the reasons for mathematical abstraction remain unchanged.

7.2. FREQUENCY CONTENT OF A SIGNAL-FOURIER TRANSFORM

One of the most fundamental characteristics of a signal is its frequency content. The frequency content of a signal provides us with a different perspective on the signal from the one we get by viewing the signal in the time domain. The frequency content of an analog signal $x(t)$ is obtained by taking its Fourier transform as follows:

$$F\{x(t)\} \equiv X(f) \equiv \int_{-\infty}^{\infty} x(t)\exp(-j2\pi ft)dt \qquad (7.3)$$

The inverse Fourier transform is given by

$$x(t) = \int_{-\infty}^{\infty} X(f) \exp(j2\pi ft) df \qquad (7.4)$$

Following usual convention, the lowercase x is used to represent the time-domain function, and the uppercase X is used for the frequency-domain function. The two functions $x(t)$ and $X(f)$ form a Fourier transform pair; this is represented symbolically by $x(t) \Leftrightarrow X(f)$. Although functionally, $x(t)$ and $X(f)$ may appear to be quite different, they nevertheless contain exactly the same information. $x(t)$ contains the information in the time domain whereas $X(f)$ contains the information in the frequency domain. Knowing one of these functions, one can get the other through the appropriate transform relationship given above.

The magnitude of $X(f)$ at any particular value f is related to the sinusoid of frequency f that is present in $x(t)$. Therefore, a plot of $|X(f)|$ shows at a glance all the frequencies that constitute $x(t)$. This plot is called the *frequency spectrum of* $x(t)$. The following examples illustrate how Eq. (7.3) is applied to some signals commonly encountered in signal processing applications.

Example 7.1: Decaying Exponential Signal.

Consider $x(t) = A \exp(-at)$, $t \geq 0$, where A and a are both positive constants. Then

$$X(f) = \int_{0}^{\infty} A \exp(-at) \exp(-j2\pi ft) dt = \frac{A}{a + j2\pi f}$$

From the frequency spectrum of $x(t)$ shown in Figure 7.1, we see that the amplitudes of the lower frequencies are uniformly greater than those of the higher frequencies. This shows that the decaying exponential $x(t)$ contains predominantly low frequencies. A signal such as this whose frequency spectrum uniformly decreases with frequency f is called a *low-pass signal.*

Example 7.2: Impulse Function $\delta(t)$.

The impulse function (or delta function) $\delta(t)$ is defined in terms of its properties as follows:

$$\delta(t) = 0 \quad \text{for all } t \neq 0 \quad \text{and} \quad \int_{-\infty}^{\infty} y(\tau)\delta(t - \tau)d\tau = y(\tau)\big|_{\tau=t} = y(t)$$

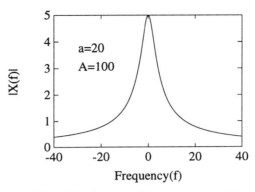

Figure 7.1. Spectrum of decaying exponential.

In words, as the variable of integration, τ, scans across all values of τ lying between $-\infty$ and $+\infty$, the impulse function is zero everywhere except at that value of τ, which makes the argument of $\delta(t-\tau)$ equal to zero. When the argument is zero at $\tau = t$, the value of the integral is given by the value of the function y at that point. The impulse function $\delta(t-\tau)$, in effect, picks off the value of $y(\tau)$ at $\tau = t$ during the integration process. From these defining properties, it follows that the Fourier transform of $\delta(t)$ is given by

$$F\{\delta(t)\} = \int_{-\infty}^{\infty} \delta(t)\exp(-j2\pi ft)dt = \exp(-j2\pi f0) = 1$$

This result states that $\delta(t)$ has a flat frequency spectrum extending over all frequencies. No matter how high the frequency is, the spectrum never falls off towards zero. No physical signal can have this property because the spectrum of all real signals eventually falls off toward zero for sufficiently high frequencies. Nevertheless, $\delta(t)$ is of immense utility in signal processing applications, particularly in analyzing the impact of the sampling process on analog signals.

7.2.1. Low-Pass, Band-Pass, and High-Pass Signals

We have seen that low-pass signals have a frequency spectrum that uniformly decreases as f increases from zero. Some signals, on the other hand, contain frequencies lying in a band centered about some frequency f_0. The frequency spectrum of these signals has a single peak centered at f_0. For frequencies lying above or below f_0, the frequency spectrum falls off toward zero. These signals are called *band-pass signals*. Finally, signals that contain predominantly high frequencies are called *high-pass signals*.

7.2.2.   Properties of the Fourier Transform

Although the Fourier transform has many properties, only a few of the more important ones are discussed here. The first of these is the relationship between multiplication in the frequency domain and convolution in the time domain. This relationship is of central importance in understanding how filters are designed to eliminate unwanted frequency components from a signal.

Property 7.1: Time-Domain Convolution.

Consider the following Fourier transform pairs:

$$x(t) \Leftrightarrow X(f) \qquad \text{and} \qquad y(t) \Leftrightarrow Y(f)$$

The time-domain convolution property states that

$$X(f)Y(f) \Leftrightarrow \int_{-\infty}^{\infty} x(\tau)y(t-\tau)d\tau$$

In words, it says that when the Fourier transforms of two time functions are multiplied in the frequency (f) domain, the result is equivalent to *convolving* the time functions in the time domain.

To prove this property, we proceed formally to evaluate the inverse Fourier transform of $X(f)Y(f)$ using Eq. (7.4). If $z(t)$ is the inverse transform, then

$$z(t) = \int_{-\infty}^{\infty} X(f)Y(f)\exp(j2\pi ft)df$$

Now $X(f) = \int_{-\infty}^{\infty} x(\tau)\exp(-j2\pi f\tau)d\tau$. Therefore, substituting for $X(f)$, we obtain

$$z(t) = \int_{-\infty}^{\infty} \left[\int_{-\infty}^{\infty} x(\tau)\exp(-j2\pi f\tau)d\tau \right] Y(f)\exp(j2\pi ft)df$$

Interchanging the order of integration and combining terms yields

$$z(t) = \int_{-\infty}^{\infty} x(\tau)\left[\int_{-\infty}^{\infty} Y(f)\exp[j2\pi f(t-\tau)df \right]d\tau = \int_{-\infty}^{\infty} x(\tau)y(t-\tau)d\tau$$

Hence $X(f)Y(f) \Leftrightarrow \int_{-\infty}^{\infty} x(\tau)y(t-\tau)d\tau \equiv x(t)\star y(t)$. This integral, which is of special importance in both transform theory and in signal processing is known as the *convolution* integral and is usually represented by $x(t)\star y(t)$. Property 7.1 relates an operation carried out in the frequency domain to its equivalent operation carried out in the time domain. Although both operations are equivalent, in some situations one is easier to carry out than the other.

The next property has important implications for sampling analog signals.

Property 7.2: Frequency-Domain Convolution.

Consider the following Fourier transform pairs

$$x(t) \Leftrightarrow X(f) \quad \text{and} \quad y(t) \Leftrightarrow Y(f)$$

Then the frequency-domain convolution property states that

$$x(t)y(t) \Leftrightarrow \int_{-\infty}^{\infty} X(u)Y(f-u)du = X(f) \star Y(f)$$

In words, this says that when two functions are multiplied in the time domain, the result is equivalent to *convolving* the Fourier transforms of these functions in the frequency domain.

To prove this property, we proceed formally to evaluate the Fourier transform of $x(t)y(t)$. Since the procedure is similar to that used in proving Property 7.1, several steps are omitted.

If $R(f)$ is the transform of $x(t)y(t)$, then $R(f) = \int_{-\infty}^{\infty} x(t)y(t)\exp(-j2\pi ft)dt$. Now $y(t) = \int_{-\infty}^{\infty} Y(u)\exp(j2\pi ut)du$. Substituting for $y(t)$, reversing the order of integration and combining terms, we get

$$R(f) = \int_{-\infty}^{\infty} Y(u)\left[\int_{-\infty}^{\infty} x(t)\exp[-j2\pi(f-u)t]dt\right]du$$

$$= \int_{-\infty}^{\infty} Y(u)X(f-u)du \equiv X(f)\star Y(f)$$

The next property is used in a later section to establish the criteria for distortionless filter design.

Property 7.3: Time-Delay Property.

If $x(t) \Leftrightarrow X(f)$, then $x(t-T) \Leftrightarrow \exp(-j2\pi fT)X(f)$.

Proof:

$$F\{x(t-T)\} = \int_{-\infty}^{\infty} x(t-T)\exp(-j2\pi ft)dt.$$

Making the change of variable $\tau = (t-T)$, we get

$$F\{x(t-T)\} = \int_{-\infty}^{\infty} x(\tau)\exp[-j2\pi f(\tau+T)]d\tau = \exp(-j2\pi f T)X(f)$$

Observe that a delay of T in the time domain corresponds to a phase-shift term $\exp(-j2\pi fT)$ in the frequency domain.

Example 7.3: Time-domain convolution.

The convolution integral is of sufficient importance in signal processing that it is worthwhile to review carefully the steps followed in its evaluation. Consider the two square waves $x(u)$ and $y(u)$ having different amplitudes (A and B) and lengths (T_1 and T_2, $T_1 > T_2$) shown in Figure 7.2. We wish to evaluate the convolution integral

$$z(t) = \int_{-\infty}^{\infty} x(u)y(t-u)du$$

The first step is to find $y(-u)$. This is accomplished by forming the mirror image of (or flipping) $y(u)$ about the $u = 0$ axis. Next, $y(t - u)$ is found by shifting $y(-u)$ to the right (left) by a distance t if t is positive (negative). The limits of integration are then selected on the basis of those values of the variable of integration u, which yield a nonzero value for the product $x(u)y(t-u)$. As a visual help with this selection process, the appropriate intervals of integration are outlined by the cross-hatched regions in Figure 7.2.

For $0 < t < T_2$, the convolution integral over the crosshatched region in Figure 7.2c is

$$z(t) = \int_{u=0}^{t} x(u)y(t-u)du = \int_{u=0}^{t} AB \; du = ABt, \qquad 0 < t < T_2$$

Proceeding similarly for $T_2 < t < T_1$ in Figure 7.2d, the convolution integral is

$$z(t) = \int_{u=t-T_2}^{t} x(u)y(t-u)du = AB[t - (t - T_2)] = ABT_2, \qquad T_2 < t < T_1$$

Finally, for $T_1 < t < T_1 + T_2$ in Figure 7.2e

$$z(t) = \int_{t-T_2}^{T_1} x(u)y(t-u)du = AB[T_1 - (t - T_2)]$$

$$= AB(T_1 + T_2 - t), \qquad T_1 < t < T_1 + T_2$$

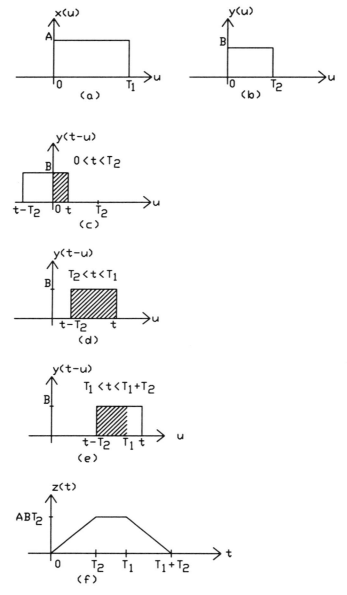

Figure 7.2. $(a),(b)$ $x(u)$ and $y(u)$; $(c),(d),(e)$ intermediate steps; (f) final result.

The convolution integral $z(t)$ is zero for all other values of t. Figure 7.2f shows the complete function $z(t)$.

Observe that $z(t)$ is longer than either of the individual functions x or y. Indeed, $z(t)$ has a length equal to the sum of the lengths of x and y. This is always true of the convolution operator; the convolution of two arbitrary functions results in a function whose length is given by the sum of the lengths of the two original functions.

Having looked at analog signals and their representation in both the time and frequency domains, we now consider how these signals are filtered.

7.3. CONTINUOUS-TIME-DOMAIN FILTERING

Figure 7.3 shows a signal $x(t)$ at the input to a filter having an impulse response $h(t)$. The output signal is $y(t)$. The impulse response of a filter is defined to be the output observed when the input is the delta function $\delta(t)$. The relationship between $x(t)$, $h(t)$, and $y(t)$ is most easily grasped initially in the frequency domain. Defining the Fourier transforms

$$x(t) \Leftrightarrow X(f) \qquad h(t) \Leftrightarrow H(f) \qquad y(t) \Leftrightarrow Y(f)$$

then $X(f)$, $H(f)$, and $Y(f)$ are related as follows in the frequency domain:

$$Y(f) = H(f)X(f)$$

The basis for this relationship lies in certain (generally straightforward) properties of linear systems that are not addressed here because of space limitations. The Fourier transform of the impulse response, $H(f)$, is known as the *transfer function* of the filter. Thus we see that the transform of the filter output is given by the product of the transform of the input signal and the filter transfer function.

With the preceding relationship established in the frequency domain, the corresponding relationship in the time domain follows immediately from Property 7.1:

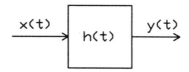

Figure 7.3. Filter with input $x(t)$ and output $y(t)$.

$$Y(f) = H(f)X(f) \Leftrightarrow y(t) = \int_{-\infty}^{\infty} x(\tau)h(t-\tau)d\tau = \int_{-\infty}^{\infty} h(\tau)x(t-\tau)d\tau$$

This is a general result. The output of a filter in the time domain is given by the convolution of the input and the impulse response functions.

We now examine a typical filtering application involving a square wave at the input to a low-pass filter. The impact of the filter on the input in both the time and frequency domains will be evident from the filter output.

Example 7.4: Response of a Low-Pass Filter to a Square Wave.

Consider a filter with the impulse response $h(t) = \exp(-t)$, $t > 0$ shown in Figure 7.4a. This is a low-pass filter because its impulse response $h(t)$, a decaying exponential, is a low-pass function (see Figure 7.1). We also note that $h(t) = 0$ for $t < 0$. A filter whose impulse response is zero for $t < 0$ is known as a *causal* filter. The term *causal* implies that no output can be obtained from the filter prior to the application of an input. Clearly, all physical filters fall into this category.

Continuing with the example, the input $x(t)$ to the filter is a single pulse of amplitude A and period T, as shown in Figure 7.4b. Following the steps outlined in Figure 7.2, we find that the output $y(t)$ in Figure 7.4c is

$$y(t) = \int_{\tau=0}^{t} A \exp(-\tau)d\tau = A[1-\exp(-t)], \qquad 0 < t < T$$

$$= \int_{\tau=t-T}^{t} A \exp(-\tau)d\tau = A[1-\exp(-T)]\exp[-(t-T)], \qquad T < t$$

We observe that $y(t)$ is considerably different from the input $x(t)$. For example, while $x(t)$ is of length T, $y(t)$ is longer, extending (theoretically) for all time. Actually, what happens in reality is that after a finite length of time, the output becomes so small that sensors are incapable of measuring any further changes in $y(t)$.

We also observe that the abrupt changes in $x(t)$ at $t = 0$ and again at $t = T$ are noticeably absent in $y(t)$. The response of $y(t)$ is relatively sluggish in each case. In order for $x(t)$ to undergo such abrupt changes, Fourier analysis indicates that it must contain high frequencies. By contrast, the slow response of $y(t)$ indicates that it is a predominantly low-frequency signal. Of course, this is precisely the effect we would anticipate from a low-pass filter. As $x(t)$ passes through the filter, the low-pass filter removes the high-frequency components from $x(t)$ and only allows the low frequencies to pass through into $y(t)$.

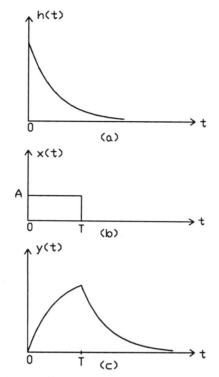

Figure 7.4. Low-pass filtering: (*a*) impulse response; (*b*) input; (*c*) output.

7.3.1. Distortionless Filtering

In some situations this distortion of the input signal by the filter is undesirable. Certain signal processing applications require the filter output to be a delayed version of the input signal with at most an amplitude change. For such a (distortionless) filter, the input $x(t)$ and output $y(t)$ are related as follows: $y(t) = Gx(t - T)$, where G is the gain of the filter and T is the time delay. The transfer function $H(f)$ of a filter with this response is obtained by taking the Fourier transform of both sides. Using Property 7.3, we get $Y(f) = G \exp(-j2\pi fT)X(f)$ and therefore $H(f) = Y(f)/X(f) = G \exp(-j2\pi fT)$. Observe that the phase response of the filter, given by the exponential term, is linear in frequency. This is an essential requirement. A distortionless filter must have a linear phase response. This requirement of linear phase places restrictions on the type of filter that can be used for distortionless filtering. Usually linear-phase filtering calls for the use of a feedforward filter structure. This will be discussed in more detail later.

Having looked briefly at continuous-time signals and some of their properties, we now consider signals in the discrete-time domain.

7.4. DISCRETE-TIME SIGNALS

The past 20 years have witnessed an explosive growth in discrete-time signal processing [referred to as a digital signal processing of (DSP) hereafter] to the point where in some applications it has completely replaced traditional continuous-time (analog) signal processing. It is worthwhile, therefore, to review some of the factors that make digital signal processing so attractive.

7.4.1. Advantages of Digital Signal Processing

Infinite Repeatability. Because digital signal processing is based on binary numbers stored in registers, no matter how often a given series of calculations is repeated, the answer is always the same. Analog computers, which use amplifiers, voltage supplies, and other components for their operation, are subject to drift over a period of time and hence may not always yield exactly the same result for a given series of computations.

Dynamic Range. By using scientific notation to represent a number, an enormous range of numerical values can be stored and processed digitally. In analog equipment, the range of numerical values, by comparison, is extremely limited. Usually the range of values is determined by the equipment power supply voltage or by the amplifier saturation voltage.

The development in the mid-1960s of a computationally efficient technique for generating the Fourier transform of a digital signal [the Fast Fourier transform (FFT)] combined with increasingly powerful and miniaturized digital processing hardware gave an enormous additional impetus to digital signal processing techniques.

While digital signal processing has some very attractive features, it is also subject to some important limitations as follows:

Many physical signals are analog in nature and consequently must first be sampled and converted into a digital signal in order to be stored and processed on a digital computer. This sampling process, which is known as *analog-to-digital* (A/D) *conversion,* is time-consuming and the hardware required can be expensive. The hardware not only performs the physical sampling of the analog signal but must also convert the sampled value into its closest binary representation. This latter step involves stepping through a series of discrete binary values, comparing each one in turn with the input value, until the closest match is found. This closest match is next stored in a register, and only then is the hardware available to take another sample of the analog signal. Clearly this is a time-consuming operation, particularly if many binary values must be sifted through.

The numerical calculations required for DSP are also time-consuming. Together with the A/D conversion time, they place limitations on the frequency ranges that can be processed digitally in real time. However, with the ever-increasing speed capabilities of new hardware, the boundaries of these limitations are constantly being rolled back.

As part of the sampling process, we shall see shortly that generally undesirable aliasing effects are always present and must be properly accounted for in the sampled signal.

Because of its central importance, we now give a detailed look at the process of sampling an analog signal.

7.4.2. Sampling an Analog Signal

Consider Figure 7.5a, which shows an analog signal $x(t)$. This signal is to be sampled every T seconds to form the sampled signal $x_s(t)$. Ignoring Figure 7.5b for the present, we observe from Figure 7.5c that $x_s(t)$ consists of the instantaneous values of $x(t)$ at each sample instant kT. The values of $x(t)$ in the time interval *between* the sampling instants are lost. However, if certain requirements as stipulated by the sampling theorem (discussed later) are satisfied, then the information lost in those values can be retrieved from $x_s(t)$.

Each sampled value is converted into its closest binary representation and then stored in a register. Usually there is some difference between these two values. The magnitude of the difference depends on the number of binary digits or bits used to represent the sampled signal. The larger the "wordlength," the smaller the difference. However, increasing the wordlength also increases the cost and complexity of the A/D hardware. In addition, it increases the time required for each conversion because more binary values must be searched through in order to find the one closest to the measured value. Finally, each conversion must be completed within the sample interval T in order to be ready to take the next sample. Compounding the problem is the ceiling on the size of T, which is imposed by the sampling theorem. Consequently, a tradeoff must invariably be made between the sample interval T, wordlength, processing needs, and cost considerations in selecting an A/D converter for any particular application.

The loss of precision in A/D conversion, known as *quantization effects,* is usually analyzed in statistical terms. A uniform probability density function between $\pm\Delta/2$, where Δ is the quantization step size, is typically used to model the conversion errors. These errors cause a lowering of the signal-to-noise ratio (SNR) in the sampled data. Formulas are derived in the references that given the minimum wordlength necessary to preserve a specific SNR in the sampled data.

Having looked at some of the hardware and cost issues associated with the sam-

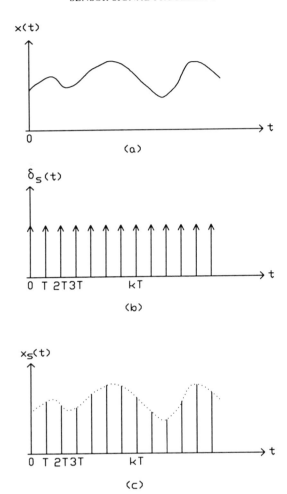

Figure 7.5. Sampling an analog signal $x(t)$: (a) analog signal $x(t)$; (b) infinite impulse train; (c) sampled signal $x_s(t)$.

pling process, we now turn our attention to the underlying mathematical issues embodied in the sampling theorem.

7.4.3. The Sampling Theorem

The most important mathematical issues relating to sampling an analog signal are addressed by the sampling theorem. To develop this theorem, suppose an analog

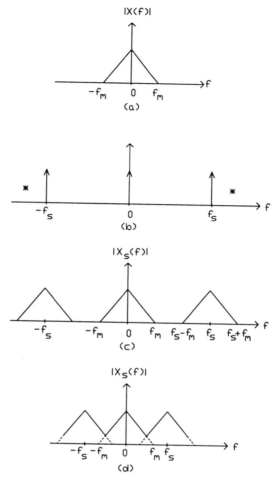

Figure 7.6. (*a*) Original spectrum; (*b*) transform of infinite impulse train; (*c*) alias-free spectrum, (*d*) spectrum with aliasing.

signal $x(t)$ having the frequency spectrum $|X(f)|$ in Figure 7.6a is to be sampled every T seconds. We observe that $x(t)$ contains no frequencies above f_m. Referring to Figure 7.5c, we observe that the sampling process in effect picks off the values of $x(t)$ at multiples of the sampling interval T. But we have already seen that the impulse function mathematically picks off the value of a function when its argument is zero. Consequently, if we *multiply* $x(t)$ by the infinite impulse train $\delta_s(t)$

shown in Figure 7.5b, we will generate a series of samples of $x(t)$ every T seconds. If $x_s(t)$ is the sampled function, then

$$x_s(t) = x(t)\delta_s(t) = \sum_{n=0}^{\infty} x(t)\delta(t - nT)$$

Clearly $x_s(t)$ is zero at all values of t except for $t = 0, T, 2T, 3T, \ldots$, where it assumes the values of $x(t)$.

The crucial mathematical information about sampling that we seek is to be found by taking the Fourier transform of $x_s(t)$. Since $x_s(t)$ is the product of two time functions, Property 7.2 states that its transform is given by the convolution of the Fourier transforms of the two functions. The transform of $x(t)$ is $X(f)$ in Figure 7.6a. The transform of $\delta_s(t)$, which is not developed here because of space limitations is shown in Figure 7.6b. Observe that it consists of an infinite impulse train (three of which are shown), separated in the frequency domain by the sampling frequency $f_s = 1/T$.

Convolving $X(f)$ with this infinite impulse train results in a spectrum $X_s(f)$ in which $|X(f)|$ is periodically repeated an infinite number of times, centered on the impulse locations. In Figure 7.6c, which shows the situation in the vicinity of $f = 0$, we observe that there is no overlap between the repeated spectra provided $f_s - f_m \geq f_m$. Avoidance of overlap is crucial if we are to avoid distortion and loss of information in the sampling process. When no overlap takes place, the original analog signal $x(t)$ can be recovered by low-pass filtering $x_s(t)$. The low-pass filter allows frequencies below f_m, namely, $X(f)$, to pass through and eliminates those above. Consequently $x(t)$ can be recovered at the filter output.

On the other hand if $f_s - f_m \leq f_m$, then the situation in Figure 7.6d exists where there is overlap of adjacent spectra. This distortion, which is known as *aliasing*, makes it impossible to fully recover $x(t)$ from the data and loss of information has occurred.

We conclude therefore that no information is lost in the sampling process if and only if $f_s = 1/T \geq 2f_m$. This is the essence of the celebrated sampling theorem. In words, it states that if a signal contains frequencies no higher than f_m, then that signal must be sampled at a rate *not less than* $2f_m$ in order to avoid loss of information. This minimum sampling frequency is known as the *Nyquist rate*. In terms of the sampling interval T, the requirement is $T \leq 1/(2f_m)$.

7.4.3.1. Hardware Implications of the Sampling Theorem

The sampling theorem imposes a new and important constraint when selecting an A/D converter. The sampling rate of the A/D hardware must be at least twice that of the maximum frequency contained in the signal. Since a high sampling rate and long wordlength have conflicting requirements, manufacturers of commercial A/D products expend considerable resources to develop new technology to accommodate both of these features. Current boards for PCs are capable of sampling rates of 100 kHz on several channels. However, these numbers are not as clearcut as they initially appear to be. They must be interpreted in the context of several application specific circumstances such as the processing speed of the PC itself, whether the board is programmed in the faster executing assembly language or in a slower high-level language such as BASIC (Beginner's All-purpose Symbolic Instruction Code), and whether there is available fast-access data storage capabilities. If the application calls for real-time signal processing of the sampled data, then the execution time for this must also be accommodated within the sample interval T. Overshadowing all of these considerations is the immutable ceiling on T itself set forth by the sampling theorem, namely, that $T \leq 1/(2f_m)$.

7.4.3.2. Antialiasing Filter

Some A/D boards incorporate an antialiasing analog filter. The analog signal is passed through this filter prior to the sampling process. This removes the signal frequencies lying above the filter cutoff frequency. The cutoff frequency can usually be set to one of several values by means of appropriate resistor connections. This prefiltering of the signal is used when we either are not interested in signal frequencies lying above the filter cutoff value or wish to eliminate any possibility of aliasing by removing those frequencies above one-half of the sampling frequency.

Now that the analog signal has been sampled and stored, we look at techniques for processing the signal digitally. As is the case with analog signals, it is also important to know the frequency content of a digital signal. That issue is addressed first.

7.5. FREQUENCY CONTENT OF A DIGITAL SIGNAL

In practical applications where a signal is sampled and stored on a computer, we seldom know the mathematical function $x(t)$ that gave rise to the data. Consequently, we cannot use Eq. (7.3) to calculate a closed-form solution for the Fourier transform of $x(t)$. In the absence of this information, we resort to numerical techniques to calculate the (approximate) Fourier transform of the signal directly from the sampled values.

7.5.1. Discrete-Time Fourier Transform

To develop this technique, suppose we have N samples of some signal $x(t)$. If the sampling interval is T, then, by definition

$$F\{x_s(t)\} = F\left\{\sum_{n=0}^{N-1} x(t)\delta(t-nT)\right\}$$

$$= \int_{t=0}^{\infty} \sum_{n=0}^{N-1} x(t)\delta(t-nT)\exp(-j2\pi ft)dt$$

$$= \sum_{n=0}^{N-1} x(nT)\exp(-jn\omega T)$$

$$= X(\omega T) = X(\Theta)$$

where $\Theta = \omega T = 2\pi fT$. This transform, which is known as the *discrete-time Fourier transform* (DTFT), is continuous in frequency and therefore cannot be calculated on a digital computer. Nevertheless, it is an important intermediate result that merits several observations. We observe that $X(\Theta) = X(\Theta \pm m2\pi)$, for any integer m. Thus, $X(\Theta)$ repeats every 2π radians in the Θ domain. Also, $X(2\pi - \Theta) = X(\Theta)^*$, where * denotes complex conjugate. Consequently, $|X(\Theta)|$ has mirror-image symmetry about $\Theta = \pi$. Figure 7.7 shows these properties for an $X(\Theta)$ that has no frequencies higher than f_m and is triangular in the region $(-\pi, \pi)$. The sampling frequency $f_s = 2f_m$. The horizontal axis is calibrated in terms of both Θ and f. We see that the maximum signal frequency f_m occurs at $\Theta = \pi$ when $f_s = 2f_m$ and the Nyquist frequency appears at $\Theta = 2\pi$.

We now derive an important property of the discrete-time Fourier transform that is analogous to Property 7.3 of the Fourier transform.

7.5.1.1. Time-Delay Property of the DTFT

Suppose that the sequence $x(n)$ having DTFT $X(\Theta)$ is delayed k sample intervals to form the shifted sequence $x(n-k)$. Then the DTFT of $x(n-k)$ is related to that of the unshifted sequence in the following manner: DTFT$\{x(n-k)\} = \exp(-j2\pi fkT)X(\Theta)$. We prove this by formally taking the DTFT as follows:

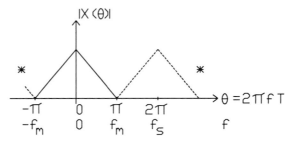

Figure 7.7. Discrete-time Fourier transform.

$$\sum_{n=k}^{N-1+k} x[(n-k)T]\exp(-j2\pi f n T)$$

Making the change of index $m = (n-k)$, and substituting, results in

$$\sum_{m=0}^{N-1} x(mT)\exp[-j2\pi f(m+k)T] = \exp(-j2\pi f k T)\sum_{m=0}^{N-1} x(mT)\exp(-j2\pi f m T)$$

$$= \exp(-j2\pi f k T)X(\theta)$$

and the property is proved. Observe that a delay of k sample intervals in the time domain corresponds to the phase shift term $\exp(-j2\pi f k T)$ in the frequency domain. This property is used later to design distortionless digital filters.

7.6. DISCRETE FOURIER TRANSFORM

Since $X(\Theta)$ is continuous and periodic, we can approximate it by computing N samples uniformly spaced in the interval $\Theta = (0,2\pi)$. Denoting these N samples as $X(\Theta_k) \equiv X(k)$ where $\Theta_k = k(2\pi/N)$, $k = 0,1,2,\ldots,(N-1)$, the transform that emerges when we substitute for Θ_k is known as the *discrete Fourier transform* (DFT), and is given by

$$X(k) = \sum_{n=0}^{N-1} x(nT)\exp\frac{-j2\pi nk}{N}, \qquad k = 0,1,2,3,\ldots,(N-1) \qquad (7.5)$$

The frequency corresponding to a given value $k \le N/2$, obtained by solving for f_k in $\Theta_k = 2\pi f_k T = k(2\pi/N)$, is $f_k = k/(NT)$. These k values are sometimes referred to as "bins."

7.6.1. Properties of the Discrete Fourier Transform

Since it is a sampled version of $X(\Theta)$, the values $X(k)$ exhibit the same periodicity and mirror-image properties of $X(\Theta)$. For example, $X(k) = X(k \pm mN)$, m an integer, and $X(N-k) = X(k)^*$, where * denotes complex conjugate. These can be easily verified by substitution in Eq. (7.5).

The frequency *resolution* of the DFT is given by the quantity $1/(NT)$. This is the smallest difference in frequency that the algorithm can resolve. Clearly, increased resolution is achieved at the expense of a longer set of data. If N is an even number, then $k = N/2$ corresponds to the maximum signal frequency, provided $f_s = 2f_m$.

The original samples $x(nT)$ are recovered from the transformed values $X(k)$ by the following inverse discrete Fourier transform (IDFT):

$$x(n) = \frac{1}{N} \sum_{k=0}^{N-1} X(k) \exp\frac{j2\pi nk}{N}, \qquad n = 0, 1, 2, 3, \ldots, (N-1) \qquad (7.6)$$

It is important to note that the original $x(n)$ values [the T in the $x(nT)$ has been dropped for convenience] can be recovered *exactly* from the $X(k)$ values. This makes Eqs. (7.5) and (7.6) an exact transform pair. No information is lost in going from one to the other except in the following sense. It is easily shown by substitution in Eq. (7.6) that $x(n) = x(n \pm mN)$ for any integer m. This implies that the sequence $x(n)$ obtained from Eq. (7.6) is of infinite length and *always* periodic in n with period N even if the original data set $x(n)$ of length N is not periodic. This periodicity of $x(n)$ has wide-ranging implications for digital signal processing procedures when implemented using the DFT.

Example 7.5: DFT of an Eight-Point Sequence.

We now consider an example of the DFT for eight-point sequence $x(n)$ shown in Figure 7.8*a*. Since the coefficients $X(k)$ are generally complex numbers having both a magnitude and a phase, they are plotted separately in Figures 7.8*b* and 7.8*c*, respectively. Note the even (mirror) symmetry of $|X(k)|$ and the odd symmetry of the phase about $k = N/2$. Because this symmetry is always present, only the values between 0 and $N/2$ need be plotted. The other values correspond to the negative frequencies in the transform. The numerical values are summarized in Table 7.1. The frequencies corresponding to $X(k)$ cannot be determined because the sampling interval T is not given. Without the associated sample interval, a sampled signal $x(n)$ and its DFT are a collection of numbers that represents incomplete information about the frequency content of $x(n)$.

Example 7.6: Effect of Frequency Resolution on the DFT.

To examine the effect of frequency resolution on the DFT, suppose that the 49-Hz signal $x(t) = \sin(98\pi t)$ is sampled every $T = 0.005$ seconds ($f_s = 200$ Hz). A total of 200 samples are taken, making the frequency resolution $\Delta f = 1$ Hz. The DFT of the resulting sequence is shown in Figure 7.9*a*. The plot, expanded about the frequency 50 Hz, shows a peak at frequency 49 Hz. All other plot values are zero. This indicates a 49-Hz signal component in the data. Note that a peak also occurs at bin number 151 (not shown in the figure). This peak does *not* mean that $x(t)$ contains a 151-Hz component. Rather, the bins above $N/2$ (= 100 here), represent the negative frequencies in the Fourier transform, which have been shifted to the right an amount f_s by the aliasing action depicted in Figure 7.6*c*. Consequently, the true frequency associated with bin 151 is $(151 - 200) = -49$ Hz. Observe that the numerical value of the peak is 100, which does *not* correspond to the amplitude of the sinusoid (which is unity). This is a general property of the DFT. The numeri-

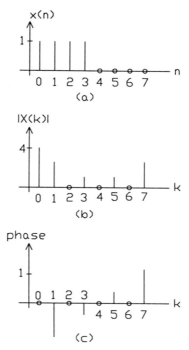

Figure 7.8. DFT of an eight-point sequence: (a) input sequence; (b) magnitude; (c) phase (radians).

cal values of $X(k)$ are *related* to the amplitudes of the frequencies present in the sampled signal, but in general are not *equal* to the amplitudes. In the present context where the signal has a single frequency, and that frequency corresponds to one of the DFT bins, it can be shown that the relationship between the value of the peak, the signal amplitude A and the number of samples N is given by peak $= AN/2$.

Continuing with the example, suppose that, instead of generating 200 samples,

Table 7.1. Numerical Values in Fig. 7.8

n,k	$x(n)$	$\lvert X(k) \rvert$	Phase (radians)
0	1	4	0
1	1	2.6131	−1.1781
2	1	0	0
3	1	1.0824	−0.3927
4	0	0	0
5	0	1.0824	0.3927
6	0	0	0
7	0	2.6131	1.1781

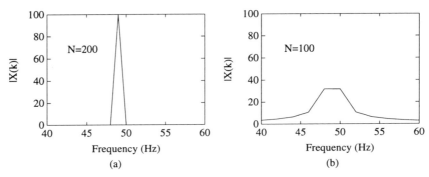

Figure 7.9. Impact of frequency resolution: (*a*) $\Delta f = 1$ Hz; (*b*) $\Delta f = 2$ Hz.

we now generate 100 samples of the 49-Hz sine wave. With $f_s = 200$ Hz as be-fore, the frequency resolution Δf is now 2 Hz. This places the signal frequency be-tween bins 24 and 25. From the DFT in Figure 7.9*b,* we see that the previous sin-gle peak has now been spread out over many bins in the vicinity of the signal frequency. This is another general characteristic of the DFT. If the frequency of a single-frequency signal does not correspond to one of the DFT bins, the resulting peak is smeared over many bins, giving the impression that several frequencies are present in the signal. These two plots demonstrate how the number of samples and the sampling interval, and consequently the frequency resolution, have a signifi-cant impact on the resulting DFT transform and why they must be carefully se-lected in any particular application.

7.6.2. The Fast Fourier Transform (FFT)

The direct evaluation of Eq. (7.5) can be very time-consuming for large values of N. When the sequence length N is an integral power of 2, a very efficient and quick method known as the *fast Fourier transform* (FFT) is used to compute the DFT. The technical details of this technique, which exploits certain properties of com-plex numbers, are discussed in many texts on digital signal processing. However, the end user of a commercial FFT software package need not be concerned about these details. Whether one uses the FFT or the DFT in Eq. (7.5), the resulting $X(k)$ values are always identical for a given data set $x(n)$.

7.6.3. Software for Computing the FFT on a PC

There is a multitude of competing PC software packages available for doing dig-ital signal processing, and all include one variety or another of the FFT algorithm.

While these FFT algorithms accept datalengths that are not integral powers of 2, their execution time for these cases may be significantly degraded. Another point to note is that some software packages cannot use zero as a register address. Therefore, the first points in the data sequences $x(n)$ and $X(k)$ start at address 1 rather than at address 0 as in Eqs. (7.5) and (7.6). In that situation, the frequency corresponding to $X(k)$ is $(k - 1)/(NT)$ rather than $k/(NT)$.

7.6.4. Time-Domain Convolution Property of the DFT

We now consider in detail the time-domain convolution property of the DFT. This is one of the most important properties of the DFT as it tells us now to relate the time-domain process of convolution to an equivalent process carried out in the frequency domain.

Suppose that the two N-point sequences x(n) and y(n) have the following discrete Fourier transforms

$$x(n) \Leftrightarrow X(k) \quad \text{and} \quad y(n) \Leftrightarrow Y(k)$$

Then the time-domain convolution property states that

$$X(k)Y(k) \Leftrightarrow \sum_{m=0}^{N-1} x(m)y(n-m)$$

To prove this, we take the IDFT of $X(k)Y(k)$. If $z(n)$ is the inverse transform, then

$$z(n) = \frac{1}{N} \sum_{k=0}^{N-1} X(k)Y(k) \exp \frac{j2\pi nk}{N}$$

Substituting the right side of Eq. (7.5) for $X(k)$, combining terms, and reversing the order of summation produces

$$z(n) = \sum_{m=0}^{N-1} x(m) \left[(1/N) \sum_{k=0}^{N-1} Y(k) \exp \frac{j2\pi(n-m)k}{N} \right] \tag{7.7}$$

$$= \sum_{m=0}^{N-1} x(m)y(n-m) = x(n) \star y(n)$$

Consequently, multiplying the DFTs in the frequency domain corresponds to the discrete convolution of the sequences in the time domain. This result is used later to implement digital filtering of a signal using the DFT.

7.7. CONVOLUTION OF DISCRETE-TIME SIGNALS

Equation (7.7) must be carefully applied when convolving two digital signals. Because of the effects of aliasing, the sequence that emerges may be either the *linear* convolution or the (generally undesirable) *circular* convolution of two signals. To understand the difference between these two, we discuss linear convolution first.

7.7.1. Linear Convolution of Two Sequences

Consider the signals $y(m)$ and $x(m)$ shown in Figures 7.10a and 7.10b, respectively, where the nonzero values are all unity. To convolve these signals using Eq. (7.7), we form $y(-m)$ in Figure 7.10c by reflecting $y(m)$ about $m = 0$. The first element of the linear convolution, $z(0)$, is obtained by summing the product of corresponding terms in $x(m)$ and $y(-m)$ that lie in the region of summation between the dashed lines. Since the two overlapping values are each equal to one, $z(0) =$

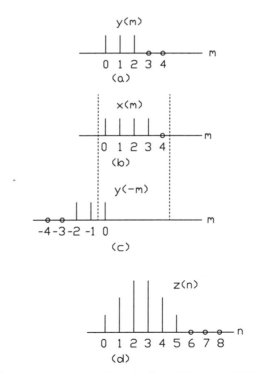

Figure 7.10. (a),(b) two input sequences; (c) $y(m)$ reflected about $m = 0$; (d) linear convolution of x and y.

1. $y(-m)$ is next shifted to the right by one unit, and the process repeated to yield $z(1) = 2$. The process of shifting $y(-m)$ to the right and summing continues until no further overlap of the two sequences takes place. The resulting function $z(n)$ in Figure 7.10d is seen to build up to a peak value of 3 and then decrease to 0.

Observe that $z(n)$ has a length equal to one less than the sum of the lengths of x and y. This is always true—the convolution of two sequences of lengths n_1 and n_2 has a length given by $(n_1 + n_2 - 1)$.

7.7.2.   Circular Convolution of Two Sequences

If we use the DFT in Eq. (7.5) to perform the preceding convolution, we would take the DFTs of $x(n)$ and of $y(n)$, multiply them, and then take the IDFT of the product. However, because the IDFT produces time sequences that have infinite periodic repetition, the IDFT in Eq. (7.6) produces the *circular* convolution of the infinite length *periodic* sequences $\hat{x}(m)$ and $\hat{y}(m)$.

To see this more clearly, consider Figure 7.11a, which shows $\hat{y}(m)$, the infinite periodic extension of $y(m)$ in Figure 7.10a. Likewise, $\hat{x}(m)$ is the infinite periodic extension of $x(m)$; $\hat{y}(-m)$ is formed by the reflection of $\hat{y}(m)$ about $m = 0$. The summation in Eq. (7.7) takes place between the two vertical dashed lines. The zero term in the circular convolution evaluates to $\hat{z}(0) = 2$. Notice that because of the repetition, additional terms now appear between the dashed lines in Figure 7.11c. These terms, which are absent in Figure 7.10c, are what produce the circular con-volution. As $\hat{y}(-m)$ is shifted to the right, the terms that exit the summation region on the right reenter the summation region on the left. Hence the term *circular con-volution.* It is left as an exercise for the reader to show that the circular convolu-tion values in Figure 7.11d are $\hat{z}(n) = (2,2,3,3,2)$. This sequence, which repeats in-definitely for further shifts to the right or left, is quite different from the linear convolution in Figure 7.10d. It is also left as an exercise for the reader to verify that the same circular convolution results are obtained using the DFT-based ap-proach.

We now look at the technique known as *zero-padding,* by which the speed of the FFT can be harnessed to linearly convolve two sequences.

7.7.3.   FFT Convolution with Zero-Padding

Suppose we wish to use the FFT to linearly convolve two sequences x_1 and x_2 hav-ing lengths n_1 and n_2, respectively. We know that the linear convolution of these two sequences has a length given by $n_1 + n_2 - 1$. The technique of zero-padding to eliminate circular convolution is as follows:

Pad the sequence x_1 with $(n_2 - 1)$ zero values to increase its length to $n_1 + n_2 - 1$. Pad the sequence x_2 with $(n_1 - 1)$ zero values to increase its length also to $n_1 + n_2 - 1$. Next, take

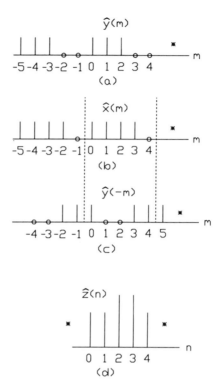

Figure 7.11. (a),(b) two input sequences; (c) $\hat{y}(m)$ reflected about $m = 0$; (d) circular convolution of $\hat{x}$ and $\hat{y}$.

the FFT of each sequence, multiply them, and take the inverse FFT. The resulting sequence is the linear convolution of x_1 and x_2. A slight complicating factor is the requirement that to utilize the full speed of the FFT, the input sequence length must be an integer power of 2. If, after padding, the sequence length $(n_1 + n_2 - 1)$ is not equal to an integer power of 2, then additional zero-padding must be applied to both input sequences x_1 and x_2 to bring their lengths up to the required power of 2 prior to taking the FFT.

The reader is invited to apply this technique to the two sequences in Figure 7.10 and verify that the result in Figure 7.10d is obtained. A word of caution. Some commercial FFT software packages automatically zero-pad the data to bring its length up to the nearest power of 2. This padding should never be relied on to eliminate circular convolution effects. The two sequences to be linearly convolved should first be sufficiently padded with zeros to eliminate circular convolution effects before they are input to the FFT software.

The problem of circular convolution can be avoided if a time-domain linear convolution algorithm is used. These time-domain algorithms, which are available in most commercial signal processing software packages, perform their calculations using the reflect, shift, and add technique described in Figure 7.10. Since the FFT is not used, the problems engendered by infinite periodic repetition do not arise. However, for very long data streams, this time-domain approach may prove unacceptably slow, making use of the FFT-based approach inevitable. Convolving long data streams will be discussed later.

7.7.4. Zero-Padding, Frequency Resolution, and the DFT

We have seen that zero-padding of the input sequence is sometimes required when using the DFT. We have also seen that the frequency resolution of the DFT is given by $1/(NT)$, where N is the sequence length and T is the sample interval. It would therefore appear that increasing the sequence length by zero-padding is a relatively painless method of improving the frequency resolution of the resulting DFT. Such is not the case, however. The frequency resolution after padding is the same as for the unpadded sequence. What the padding has accomplished for us is the ability to interpolate between the frequency bins of the unpadded data.

To appreciate this, suppose we append N zeros to a sequence $x(n)$ of length N to produce a sequence $x'(n)$ of length $2N$. The DFT of $x'(n)$ is

$$X'(k) = \sum_{n=0}^{N-1} x'(n) \exp \frac{-j2\pi nk}{2N}, \qquad k = 0,1,2, \ldots, 2N-1$$

Observe that the summation is only over the first $(N-1)$ points of $x'(n)$, as the zeros add nothing to the sum. If $X(k)$ is the N-point DFT of $x(n)$, then $X'(2k) = X(k)$, $k = 0,1,2, \ldots, (N-1)$ and the other $(N-1)$ bins of $X'(k)$ are simply interpolated values lying midway between the bins of $X(k)$. If the frequency resolution is to be improved, more data must be collected.

Having considered the representation and properties of digital signals, we now discuss how they are filtered.

7.8. INTRODUCTION TO DIGITAL FILTERS

This discussion addresses some of the most basic elements needed to understand and use digital filters. Additional details on this extensive subject can be found in the references listed at the end of this chapter.

Digital filters fall into two broad categories: feedforward and feedback. Each of these has its own characteristic structure and properties, its own advantages and disadvantages, which we now consider, starting with the feedforward filter.

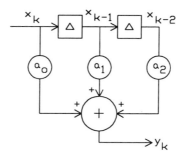

Figure 7.12. Feedforward digital filter.

7.8.1. Characteristics and Properties of Feedforward Digital Filters

Figure 7.12 illustrates the general structure of a feedforward digital filter. The present output y_k is a weighted sum of the present (x_k) and two previous (x_{k-1}, x_{k-2}) inputs. The latter are stored in shift registers represented by boxes in the figure. The Δ in each box symbolizes a unit delay. The filter output is expressed mathematically by the following difference equation:

$$y_k = a_0 x_k + a_1 x_{k-1} + a_2 x_{k-2} \qquad (7.8)$$

When the next input sample is taken, the value x_{k-1} is shifted into the right register to become x_{k-2}. Then x_k is shifted into the first register to form x_{k-1}. The next sample value becomes the new x_k, and the next output value is generated. This process is repeated for each new input sample. Because there are only two time delays (shift registers), this is a second-order filter. More sophisticated filters employ numerous storage registers. However, this simple example captures the essential elements of all of these filters.

When analyzing feedforward filters, it is often convenient to use the discrete-time Fourier transform. Applying this transform to both sides of Eq. (7.8) and using its time-delay property, we get $Y(\Theta) = a_0 X(\Theta) + a_1 \exp(-j2\pi fT)X(\Theta) + a_2 \exp(-j2\pi f2T)X(\Theta)$. The filter transfer function is

$$H(\Theta) = \frac{Y(\Theta)}{X(\Theta)} = a_0 + a_1 \exp(-j2\pi f\,T) + a_2 \exp(-j2\pi f2T) \qquad (7.9)$$

For a given set of coefficients a_0, a_1, a_2 and sample interval T, Eq. (7.9) gives the value of the filter transfer function at any particular frequency $0 \leq f \leq 1/(2T)$ where $\Theta = \omega T = 2\pi fT$. This value is, in general, a complex number. The magnitude gives the value of the filter gain, and the phase angle gives the time delay that the filter imparts to the input signal.

A particularly interesting case arises when $a_0 = a_2$. Suppose this is true; then, if we factor out the term $\exp(-j\omega T)$ from Eq. (7.9), the transfer function becomes

$$H(\Theta) = \exp(-j\omega T)[a_0 \exp(+j\omega T) + a_1 + a_0 \exp(-j\omega T)]$$
$$= \exp(-j\omega T)[a_1 + 2a_0\cos(\omega T)] \tag{7.10}$$

Observe in Eq. (7.10) that the filter phase response $\exp(-j\omega T)$ linearly increases with frequency f. We have already seen that a linear-phase response is necessary for a filter to be distortionless. This is an important property of feedforward filters, namely, that they can always be designed to have exactly linear phase. From the time-delay property of the DTFT, this phase response implies that the filter imparts a delay of one sample interval to the signal. The filter gain as a function of frequency is $[a_1 + 2a_0\cos(\Theta)]$.

Example 7.7: Design of a Feedforward Filter.

As a simple example, suppose a signal consists of the sum of two sinusoids of frequencies 50 and 60 Hz. We wish to design the filter in Figure 7.12 to eliminate the 50-Hz signal and pass the 60-Hz signal unattenuated. The sample interval is $T = 0.001$. The 50-Hz frequency translates into the angle $\Theta = 2\pi fT = 2\pi(50)(0.001) = \pi/10$ radians, and 60 Hz translates into $3\pi/25$ radians. With filter gains of 0 and 1, respectively, corresponding to these frequencies, the coefficients a_0 and a_1 are obtained by solving the following two equations:

$$0 = a_1 + 2a_0 \cos\left(\frac{\pi}{10}\right) \qquad \text{(eliminate 50 Hz)}$$

$$1 = a_1 + 2a_0 \cos\left(\frac{3\pi}{25}\right) \qquad \text{(pass 60 Hz)}$$

Using any of a variety of techniques, the solution yields $a_0 = a_2 = -23.4962$ and $a_1 = 44.6924$.

Figure 7.13 shows the filter input x_k (dash–dot), the 60-Hz component of that input (dash), and also the filter output (solid). Observe that the output is an exact replica of the input 60-Hz component. This clearly shows that the 50-Hz component of the input signal has been completely removed while the 60-Hz component is passed without any change. The output 60-Hz signal lags the input 60-Hz component by one sample interval as predicted by the filter phase response $\exp(-j2\pi fT)$.

Observe from Example 7.7 that because we have only two coefficients at our disposal ($a_2 = a_0$), it is not possible to arbitrarily specify the filter gain at a third frequency. To specify additional points, the number of coefficients (and consequently the number of shift registers) must be increased. Of course, this increases

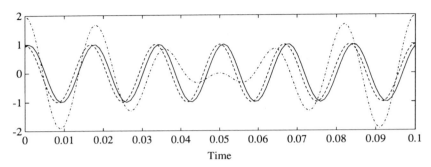

Figure 7.13. Example of feedforward digital filter design.

the filter complexity and cost and may require a lengthy filter if many frequency points need to be specified. We now consider feedforward filters in a more general setting.

7.8.2. Impulse Response of a Feedforward Digital Filter

The impulse response of a digital filter is defined as the response when the input is the Kronecker delta function. The Kronecker delta function $\delta(i - j)$ is defined to be zero for all values of $i \neq j$ and unity for $i = j$. Suppose the Kronecker delta function $\delta(i)$, which is zero everywhere except at $i = 0$, where it is unity, is applied to the filter in Figure 7.12. Assuming that both shift registers initially store zero, the filter impulse response $h(i)$ at $i = 0$ is a_0. At the next sample interval $i = 1$, the previous input of unity now appears at the output of the first storage register. The filter input is zero, and consequently the filter response $h(1)$ is a_1. This process repeats for $i = 2$ with zero at both the filter input and at the first register, while unity now appears at the second register. The response $h(2) = a_2$. The impulse response is $h(i) = (a_0, a_1, a_2)$, after which no further output is observed because the input is zero and both registers store zero. This is a general characteristic of a feedforward filter. The weighting coefficients constitute its impulse response. Since the number of weighting coefficients is always finite, the impulse response always contains a finite number of terms. Consequently, the feedforward filter is more commonly known as a *finite-impulse response* (FIR) filter. Further, no matter how large the numeric values of these coefficients may be, they always have a finite value. This gives the FIR filter an added attractive characteristic, namely, that it is always a stable filter.

Consider a general FIR filter having N terms in its impulse response $h(i)$, $i = 0, 1, 2, \ldots, (N - 1)$. From the preceding discussion we see that the output at any time j is given by

$$y(j) = h(0)x(j) + h(1)x(j-1) + h(2)x(j-2) + \cdots + h(N-1)x(j-(N-1))$$

$$= \sum_{i=0}^{N-1} h(i)x(j-i) = h(i) \star x(i) \qquad (7.11)$$

We conclude that the filter response is obtained by convolving the input with the filter impulse response. This is a general result that applies not only to FIR filters but also to feedback filters. The output signal from a digital filter is obtained by convolving the signal at the filter input with the filter impulse response. But the time-domain convolution property of the DFT tells us that the operation in Eq. (7.11) can be carried out in the frequency domain by multiplying the DFTs of the input signal and impulse response. Consequently, we can perform a digital filtering operation on a signal in either the time domain using linear convolution or in the frequency domain using the DFT. In the latter case, we must take steps to avoid circular convolution effects as discussed already.

Generalizing the example in Figure 7.12, an N-stage FIR filter with a linear-phase response is designed by incorporating into the impulse response $h(i)$ the following mirror-image symmetry: $h(i) = h(N-1-i)$, $i = 0,1,2, \ldots, (N-1)$. The ability to build this into the filter structure is what makes the FIR structure so attractive in many applications. Of course, as N increases, so also does the filter delay of $(N-1)/2$ sample intervals (assuming N is odd). However, this delay is usually deemed an acceptable price to be paid in order to achieve distortionless filtering.

In some applications the filter phase response is of secondary importance to achieving a specified amplitude response. In those cases the feedback filter structure is generally preferable to the feedforward structure, because a given amplitude response can usually be realized more economically with feedback than with feedforward. These issues are explored next.

7.8.3. Characteristics and Properties of Feedback Digital Filters

Figure 7.14 shows the general structure of a single-loop feedback digital filter. The output y_k is a function of both the current input u_k and the previous output y_{k-1}. This filter structure stands in marked contrast to the feedforward structure, where the output was a function of only the current and previous inputs. The feedback of the filter output to the input imparts to the filter a set of properties that are profoundly different from those of the feedforward structure. To appreciate this, we begin with the difference equation governing the filter: $y_k = ay_{k-1} + u_k$. If the input $u_k = \delta(k)$, then the first several terms in the impulse response for the unenergized filter ($y_{-1} = 0$) are $1, a, a^2, a^3, \ldots$. The closed-form expression for the impulse response is immediately evident; $h(n) = a^n$, $n \geq 0$.

If $|a| < 1$, then the impulse response dies out for large n and the filter is, by definition, stable. If $|a| > 1$, then $h(n)$ grows exponentially with n and the filter is un-

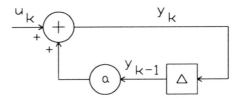

Figure 7.14. Single-loop feedback digital filter.

stable. This simple example demonstrates two important properties of a feedback digital filter. The first of these is that the filter can be unstable depending on the value of the feedback coefficient a. The second is that it has an infinite number of terms in its impulse response. For this reason, feedback digital filters are more commonly known as *infinite impulse response* (IIR) filters. Recall that with the FIR structure, the impulse response has a finite length and the filter is always stable regardless of the magnitude of the weighting coefficients.

7.8.4. Frequency Response of the Single-Loop IIR Filter

Assuming that $|a| < 1$, the DTFT of $h(n)$ is

$$H(\Theta) = \sum_{n=0}^{\infty} [a \cdot \exp(-j2\pi f T)]^n = [1 - a \cdot \exp(-j2\pi f T)]^{-1}$$

The magnitude and phase of $H(\Theta)$ are plotted in Figures 7.15*a* and 7.15*b*, respectively, for $a = 0.5$. Observe that the filter is low-pass because the magnitude uniformly decreases with frequency from an initial value of 2 at $\Theta = 0$ to a value of 0.67 at $\Theta = \pi$. Observe also that the phase is a nonlinear function of Θ. The magnitude and phase responses of this single-loop filter are inextricably mingled in the function $H(\Theta)$. There is no easy way to mathematically separate them, as was possible with the FIR filter structure. This is another important characteristic of the IIR filter. The design of an IIR filter is largely the design of the magnitude of the filter response. The filter phase response is usually nonlinear and cannot generally be specified independently of the amplitude specification. Consequently, IIR filters are used in applications where phase distortion is of no concern or where the input waveshape need not be preserved at the filter output.

Figure 7.15*a* demonstrates an important advantage of the feedback filter structure. Notice how, using a simple first-order feedback filter, we have succeeded in

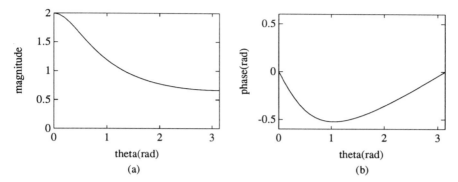

Figure 7.15. Frequency response of single-loop IIR filter: (*a*) magnitude; (*b*) phase.

defining the entire filter frequency response from $\Theta = 0$ to $\Theta = \pi$ with just one
feedback coefficient a. In general, a given frequency response can be achieved
with a lower-order filter using an IIR filter than with an FIR filter. To achieve this
frequency response using an FIR structure, one would have to specify many frequency points between $\Theta = 0$ and $\Theta = \pi$ and use a correspondingly high-order
filter.

7.8.5. Stability Requirements of Feedback Digital Filters

The requirement that $|a| < 1$ for the single-pole filter to be stable is a special case
of the generic stability requirement for all IIR digital filters, namely, that the
"poles" of the filter must lie within a circle of unit radius centered on the origin in
the Z-transform space. A more complete understanding of filter "poles" is to be
found in Z-transform theory, a full treatment of which cannot be entered into here.
However, ignoring several mathematical subtleties, the concept can be grasped by
viewing the Z-transform as the DTFT with the term $\exp(-j2\pi fnT)$ in the latter replaced by z^{-n}. With this definition, the Z-transform of the difference equation governing the feedback filter in Figure 7.14 is $Z\{y_k\} = Z\{ay_{k-1}\} + Z\{u_k\}$. Using uppercase to denote transformed quantities and recognizing that, because of its close
similarity to the DTFT, the Z-transform has a time-delay property similar to that
of the DTFT, we get $Y(z) = az^{-1}Y(z) + U(z)$. The Z-transform of the filter transfer function is $H(z) = Y(z)/U(z) = 1/(1 - az^{-1}) = z/(z - a)$. The poles of the filter are defined as the roots of the denominator polynomial of the transfer function.
A digital filter is stable only if all of its poles lie within the bounds of the unit circle. Applying this criterion to the single-loop filter, we find that it is stable only if

its pole at $z = a$ lies within the unit circle: $|a| < 1$. This agrees with what we found previously when we solved for the filter impulse response. The roots of more complex feedback filters are determined in a similar manner. The difference equation for a multiple-loop feedback filter is first determined and then Z-transformed. The transfer function is then obtained and the denominator polynomial established. The roots of this polynomial constitute the filter "poles" that must all lie within the unit circle to guarantee filter stability.

We have just seen how to get the filter transfer function from the filter difference equation. We now consider the reverse of this, namely, how to get the filter difference equation from the transfer function. This latter procedure is needed in order to implement a filter when the output from filter design software is given in terms of the polynomial coefficients in both the numerator and denominator of the transfer function.

7.8.6. Determining the Difference Equation from the Transfer Function

The transfer function of a digital filter usually involves polynomials of z^{-1} in both the numerator and denominator. The following is an example of a typical second-order filter transfer function;

$$H(z) = \frac{Y(z)}{U(z)} = \frac{s_0 + s_1 z^{-1} + s_2 z^{-2}}{r_0 + r_1 z^{-1} + r_2 z^{-2}} \qquad (7.12)$$

where $U(z)$ and $Y(z)$ are the Z-transforms of the filter input and output, respectively. The roots of the numerator polynomial are known as the *zeros* of the transfer function, for obvious reasons. Cross-multiplying in Eq. (7.12), we get $Y(z)[r_0 + r_1 z^{-1} + r_2 z^{-2}] = U(z)[s_0 + s_1 z^{-1} + s_2 z^{-2}]$. Using the time-delay property of the Z-transform, we can write this equation in its equivalent time-domain formulation: $r_0 y_k + r_1 y_{k-1} + r_2 y_{k-2} = s_0 u_k + s_1 u_{k-1} + s_2 u_{k-2}$. Solving for y_k, the difference equation that emerges is

$$\begin{aligned} y_k &= -(r_1/r_0)y_{k-1} - (r_2/r_0)y_{k-2} + (s_0/r_0)u_k + (s_1/r_0)u_{k-1} + (s_2/r_0)u_{k-2} \\ &= a_1 y_{k-1} + a_2 y_{k-2} + b_0 u_k + b_1 u_{k-1} + b_2 u_{k-2} \end{aligned} \qquad (7.13)$$

Figure 7.16 shows one possible filter realization of Eq. (7.13). Observe that it involves both a feedforward and a feedback section. If the a_i coefficients are zero, the filter is feedforward while if b_1 and b_2 are zero, it is purely feedback. Higher-order transfer functions involve additional feedforward and feedback paths; however, the technique of extracting the difference equation is the same as followed here.

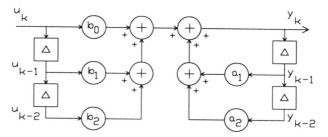

Figure 7.16. Direct-form realization of Eq. (7.13).

7.8.7. Structural Realization of the Difference Equation

Although the technique of determining the difference equation is straightforward, one must be careful in deciding which of several possible realization structures one uses to implement it. Recall that numerical values are stored in their nearest binary representation and because of the finite wordlength, there is a difference between the true value and the binary representation (known as *quantization effects*). As a result of this difference, the filter feedforward and feedback coefficients are seldom equal to the design values computed using high-precision calculations. This inaccuracy in coefficient values shifts the locations of the filter's poles and zeros, which, in turn, perturbs the filter's frequency response away from the desired value. Obviously the extent of this perturbation is a function of the wordlength; a shorter wordlength tends to exacerbate the problem. The references show that when a filter's poles and zeros are clustered together, the impact of these quantization effects can be greatly magnified. In some situations, the effect can be unacceptable. For example, if a filter design calls for the poles to be located close to the unit circle, the perturbation can cause some poles to migrate outside the circle, resulting in filter instability.

Although we generally have little control over the wordlength, there is another effect that contributes to the problem and over which we do have some control. A given difference equation can usually be realized in several different structures. For example, the direct-form structure in Figure 7.16 is not the only one for realizing Eq. (7.13). The references discuss alternate structures that can be employed such as coupled-form, cascade, and parallel. These are discussed separately for both FIR and IIR filters. It is shown that preferred structures exist by which the impact of quantization effects on the overall frequency response can be minimized for each of these filter types. Complicating the issue is whether fixed- or floating-point implementation is available. In general, high order filters should be implemented as a series of either cascade or parallel second-order sections rather than in direct form, to minimize quantization effects on the coefficients. The reader is referred to the references for more specific details on these realizations and also on the related problem of quantization effects in A/D conversion.

7.9. DIGITAL FILTER DESIGN

Many sophisticated techniques for designing digital filters have been developed in recent years. Although a detailed treatment of these methods is beyond the scope of this material, we will attempt to highlight the salient issues associated with several of them. The references provide an excellent mathematical treatment of these techniques.

7.9.1. FIR Filter Design Techniques

FIR filter design consists of two main techniques, those that specify the desired filter performance in the time domain and those that specify the desired performance in the frequency domain. We start with the former method first.

7.9.1.1. Windowing Technique of FIR Filter Design

In this technique, the filter coefficients are determined from the finite-length impulse response $h(n)$ obtained by truncating an infinite-length impulse response $h_i(n)$. The latter might be a desired response that we wish to approximate by an N-stage FIR filter by simply taking the first N terms as follows: $h(n) = h_i(n), 0 \leq n \leq (N-1)$ and $h(n) = 0, n \geq N$. Mathematically, $h(n) = h_i(n)w(n)$, where the rectangular truncation window function $w(n)$ is defined as $w(n) = 1, 0 \leq n \leq (N-1)$ and $w(n) = 0, n \geq N$. Suppose the DTFT of $h_i(n)$ and $w(n)$ are $H_i(\Theta)$ and $W(\Theta)$, respectively. Then, since $h(n)$ is obtained by multiplying $h_i(n)$ and $w(n)$ in the time domain, the transform of $h(n)$, namely $H(\Theta)$, is given by the convolution of $H_i(\Theta)$ and $W(\Theta)$ in the frequency (Θ) domain. Now we wish $H(\Theta)$ to be as close as possible to $H_i(\Theta)$ as the latter is the desired response. We have seen that the convolution operator tends to spread out and smear the input functions, and consequently, we need $W(\Theta)$ to be a narrow pulse to minimize this smearing effect. However, Fourier transform theory states that a narrow pulse in the frequency domain can be obtained only at the expense of lengthening the corresponding function, namely, $w(n)$ in this case, in the time domain. But we need to keep the window length N small to minimize the length of the FIR filter. Clearly these are conflicting requirements.

The problem is rendered more complex by the high-frequency oscillation, known as the *Gibbs effect,* arising from the abrupt changes by $w(n)$ at $n = 0$ and at $n = (N-1)$. To alleviate this effect, a series of more gradually tapered windows have been developed with which to truncate the infinite impulse response. Two of the more common ones are *Hanning* and *Hamming*. The mathematical function associated with each window type is available in most books on DSP. Once the window length N is specified, the function gives the N coefficients for that particular window. The frequency characteristics of each window are usually available in these texts. The more gradual tapering results in less high-frequency ringing but at the expense of a wider main lobe. The reader is referred to the references for a more mathematical discussion of these window functions and their role in designing FIR filters.

Ultimately, one must resort to some measure of trial and error in selecting both the impulse response length of the FIR filter to be designed and the type of window to be used in the truncation process.

7.9.1.2. Other FIR Filter Design Techniques

Another common method of FIR design is to take samples of a desired frequency response and use the DFT to invert these samples to get the filter impulse response. An important issue to be resolved in this method is the number and location of samples in the transition bands. Once again, it may be necessary to use computer-aided techniques to arrive at a final design. Note that if a distortionless filter response is desired, the frequency samples must be appropriately weighted by a linear-phase function when numerically computing the filter coefficients using the inverse DFT.

The *equiripple* approach for designing FIR filters attempts to resolve some of the approximation problems of the two methods just discussed by specifying an acceptable ripple in the frequency response in the passband and stopband regions. This ripple (or difference) between the actual and desired frequency response is minimized so that it never exceeds specified limits. Sophisticated computer-aided design algorithms have been written to design FIR filters using this approach (Parks–McClellan). These algorithms, which use mathematical optimization techniques, are usually included in current commercial DSP software packages. The mathematical details of the optimization techniques need not be understood by the end user. The filter order, frequency points, and associated filter gains are input to the algorithm, which then calculates the weighting coefficients and plots the resulting frequency response for comparison with the desired response. If the design is not satisfactory, the process is iterated on until a final design is deemed acceptable.

7.9.2. IIR Filter Design Techniques

The design of IIR filters involves extensive use of the theory of Laplace transforms and Z-transforms. In addition, the techniques build on sophisticated concepts drawn from analog filter theory. Consequently our discussion will be qualitative in nature. The references give complete mathematical details of most design methods.

Design techniques generally fall into two broad categories. In the first of these, the digital filter is designed to approximate the response of an analog filter. The second category places less reliance on analog designs and relies more on design in the digital domain. We consider the former category first.

7.9.2.1.   Designs Based on Analog Filters

These methods design a digital filter that in some way is an approximation to an analog filter. Since the field of analog filter design is very mature, many analog filters having a wide variety of frequency characteristics have already been designed. The following techniques exploit these existing analog designs to arrive at a digital filter.

In the *impulse invariance* design method, the digital filter is designed to have an impulse response that is a regularly sampled version of the impulse response of an analog filter. The design is accomplished by an exponential mapping of the analog filter poles into locations inside the unit circle. The sampling interval T is critical to this technique in order to avoid aliasing problems. Indeed, aliasing problems render this an inappropriate method for high-pass filter design.

The *bilinear transform* method is another technique based on analog filter design. It has the advantage of being free of aliasing problems. This method involves a series of steps that begin in the digital frequency domain where the filter's critical frequencies Θ_i are first selected. These frequencies are then "prewarped" into the analog frequency s-domain using the transformation $s_i = (2/T)\tan(\Theta_i/2)$. The analog filter is then designed on the basis of these s_i frequencies. The transfer function $H(s)$ of the analog filter is mapped into the discrete transfer function $H(z)$ by replacing the parameter s with the bilinear transformation $s = (2/T)(1 - z^{-1})/(1 + z^{-1})$. The resulting $H(z)$ gives the IIR filter's transfer function coefficients, which are then used to implement the filter.

It might appear to the reader that a good background in analog filter design is a necessary prerequisite for using these digital filter design techniques. Fortunately, such is not quite the case, as the references include a comprehensive discussion of several classic analog filter designs (Butterworth, Chebyshev, etc.) with appropriate closed-form design equations and worked examples. These examples enable the uninitiated to quickly master the design equations for the different analog filter types. When the analog filter is designed, the digital filter is obtained by using an appropriate transformation into the Z-domain such as the bilinear transformation, for example.

7.9.2.2.   Other IIR Filter Design Techniques

The second category involves digital filter design without reference to any analog filter. The filter frequency response is specified at several points in the frequency domain, and a mathematical optimization algorithm is then used to determine the filter coefficients. The optimization criterion usually involves minimizing a squared error cost functional based on the difference between the actual response at the specified frequencies and the desired value. The problem with a technique such as this is that a local rather than a global minimum design may result.

Several techniques exist for specifying the filter response in the time domain (without reference to any analog design). The filter is then designed using either system identification-based algorithms or least-squares methods. The reader is referred to the references for specific technical details of these methods.

7.9.3. Commercial PC Software for Digital Filter Design

Most commercial PC software packages for DSP include a suite of algorithms for both FIR and IIR digital filter design. The user inputs the filter type (FIR, Butterworth, Chebyshev, etc.), filter passband and stopband frequencies, and so on into the appropriate algorithm, which then outputs the resulting digital filter coefficients. These are usually given as numerator and denominator polynomial coefficients of the transfer function. A routine for determining filter order on the basis of frequency specifications is usually available for several filter types. In addition, an option is available to plot the actual response of the filter for immediate comparison with the desired response.

To economize on the number of input parameters, some packages use the concept of normalized frequency for their digital filter design routines. To understand how this works, recall that the transfer function $H(\Theta)$ has mirror-image symmetry about $\Theta = \pi$, specifically, $|H(\Theta)| = |H(2\pi - \Theta)|$. Consequently, only those digital frequencies lying in the range $0 < \Theta < \pi$ radians can be arbitrarily specified. Since $\Theta = 2\pi fT = 2\pi f/f_s$, this implies that frequencies (in hertz) can be arbitrarily specified only in the range $0 < f < f_s/2$, where f_s is the sampling frequency for the data. The normalized frequency f_n corresponding to a frequency f in hertz is defined as $f_n \equiv f/(f_s/2)$. Clearly the normalized frequency always has a value between zero and unity. For example, a frequency of 300 Hz when the sampling frequency is 2000 Hz would be input to the filter design routine as a normalized value of $300/1000 = 0.3$. To convert from the normalized frequency to the frequency in hertz, simply multiply the former by $(f_s/2)$. The digital frequency in radians is then obtained by multiplying this result by $2\pi/f_s$: $\Theta = 2\pi fT = f_n(f_s/2)2\pi/f_s = f_n\pi$.

Recall that the Nyquist frequency is the minimum sampling rate f_s that avoids aliasing for a low-pass signal. Some DSP packages define one-half of the sampling frequency (i.e., $f_s/2$) as the Nyquist rate when defining the input parameters to the filter design routines. Once this distinction is understood, any source of confusion should be eliminated when using the software.

Having discussed discrete-time signals and how to design digital filters, we now look at how these signals are filtered.

7.10. DIGITAL FILTERING OF SIGNALS

Suppose we wish to filter a very long data stream with a filter whose impulse response contains M points. The method one normally follows for a data stream is to collect the entire data stream and perform a linear convolution of the data with the impulse response in the time domain or in the frequency domain using the FFT with appropriate zero padding to eliminate circular convolution effects. However, for long data streams we may not wish to wait until all the data is collected, or even if we did, the extensive amount of data could present computational difficulties even for the FFT.

Two methods have been developed for filtering a long data stream while the data is still being collected. We discuss them in turn, beginning with the overlap-add method.

7.10.1 Overlap-Add Method

In the *overlap-add* method, the data stream is broken into sections of length N. Each section is linearly convolved with the filter impulse response (length M), and the resulting convolved sections are added to each other with the appropriate overlap. To use the FFT in this process, the impulse response and the section must both be padded with the appropriate number of zeros to bring the length of each up to $N + M - 1$. The two $(N + M - 1)$- point FFTs are multiplied and inverted to get the linearly convolved section of length $(N + M - 1)$. The last $(M - 1)$ points of each convolved section are overlapped with and added to the first $(M - 1)$ points of the next convolved section in order to properly account for the contiguous nature of the input data sections. This process continues for as long as the data is collected. The result is the same as collecting the entire data stream and then processing it through the filter. The linearity property of convolution is what makes this sectioning approach possible, namely $(x_1 + x_2) \star h = (x_1 \star h + x_2 \star h)$. Note that the FFT of the impulse response need be computed just once in this technique as long as the section length N remains unchanged.

7.10.2. Overlap-Save Method

In this second method, the data is again sectioned into lengths of N. Assuming $N > M$, the impulse response is zero-padded to bring its length up to N. The N-point FFTs of the sectioned data and padded impulse response are multiplied and inverted. However, since the length of these sequences is $N < N + M - 1$, the effects of circular convolution are not entirely eliminated. Indeed, the shortfall of $(M - 1)$ is the number of points which are circularly convolved. If $z(n)$ is the output, then $z(0)$ through $z(M - 2)$ are discarded because they represent circular convolution effects. The terms $z(M - 1)$ through $z(N - 1)$ are saved as they represent correct linearly convolved values. The overlap-save method consists of discarding these first $(M - 1)$ values and then concatenating the remaining points to the end of the previous convolved section. With this method overlapping takes place in the sectioned-data domain rather than in the convolved sequence domain. Additional information on these methods is available in the references.

7.10.3. IIR Filtering Using the DFT

Note that the DFT can be used to implement digital filtering using a stable IIR filter. While mathematically, the impulse response has an infinite number of terms and hence cannot be stored on a computer, in actuality it decays to zero after a finite number of terms. This is due to the finite resolution capabilities of the wordlength used to implement the filter, which regards as zero any magnitude lying below its smallest resolution about zero. The number of terms required for the impulse response to decay to zero is variable depending on the location of the filter's poles and on the wordlength. If the poles are located close to the unit circle, then it takes many more terms than when they are located close to the origin (center of the circle). In either case, the response eventually decays to zero, making it possible to use the DFT to preform IIR digital filtering

7.10.4. Zero-Phase Smoothing

In statistical studies of data one often constructs a function $y(n)$ having a single peak at some value n^* that is to be determined. However, because of random errors present in the data, the function is usually very noisy, and consequently n^* cannot be determined from merely plotting $y(n)$. To eliminate the effect of the high-frequency noise, we decide to low-pass-filter $y(n)$. In order to determine n^* when the noise is stripped away, we reason that the shape of the underlying function must not be distorted by the filtering process. Consequently we decide to use an FIR rather than an IIR filter as the former can be designed to have exactly linear phase and avoid distortion. However, the linear-phase response implies that the output sequence is shifted or delayed by the filtering process and consequently, the value n^* that we obtain from the filtered data will be in error as a result of this shift.

Zero-phase smoothing is one solution to this dilemma. This is accomplished by filtering the data with a linear-phase filter whose delay $(N-1)/2$ can always be determined ahead of time. The filtered data is then advanced by an amount equal to this delay, and the correct value n^* is then determined from the data. One of the windowing functions discussed earlier such as Hanning or Hamming usually proves suitable in this application. If the window length N is an odd number, the delay is an integer number of sample intervals; otherwise it involves a fractional part of a sample interval.

The association of zero phase with this procedure merits several comments. A causal filter cannot have zero phase. When we filter the data with the FIR filter, we are using a causal filter. When the resultant data is advanced an amount equal to the filter delay, we are in effect subtracting the delay from the filter and making it noncausal. This demonstrates that, provided we are prepared to accept a delay, we can implement noncausal filtering on stored data.

7.11. FILTERING SIGNALS CONTAINING NOISE

From the preceding paragraph we see that the introduction of noise into a signal complicates matters significantly. One must first deal with the noise before extracting the information from the underlying information-carrying signal. The field of stochastic signal processing that addresses these types of issues is extensive and cannot be discussed in detail here. Nevertheless, there are a few tools from this discipline that have proved their usefulness over time, and these are introduced in the remaining part of this chapter. We begin with the linear least-squares technique.

7.11.1. Linear Least-Squares Techniques

To see what least squares involves and where it can be employed, let us return to the zero-phase example. Suppose that the noise has been filtered from $y(n)$ and that the data in the vicinity of the peak is as depicted in Figure 7.17 (ignore the dashed line). We might be tempted to select the value $n^* = n_i$ since the latter is where the function has its maximum. However, suppose that the data represents the temperature at a specific point taken every hour and that we wish to estimate exactly when the temperature achieved its maximum. It would indeed be fortuitous if we happened to take the temperature reading at precisely the time the latter had achieved its maximum. We would be more likely to take readings at times prior to and subsequent to the peak value. In that situation, we would suspect that the maximum had not occurred at exactly n_i but at some time in the vicinity of n_i. In a situation such as this, we could use a least-squares fit to the data in the vicinity of n_i, and from that fit we can interpolate to find a closer estimate of when the peak actually occurred. Since we are looking for a peak, a second-order polynomial least-squares fit to the data in the vicinity of n_i would be an initial choice (dashed line in Figure 7.17).

The following derivation of the least-squares fit is generalized slightly from the preceding discussion to make it easier to apply in a wide variety of situations. Suppose we have a collection of data points $(y_j, x_j), j = 1, 2, \ldots, N$, where x_j is the independent variable and y_j is the dependent variable or observed value. In the context of the preceding example, x_j would be the time at which the temperature was taken and y_j would be the observed value. The second-order fit we seek is given by

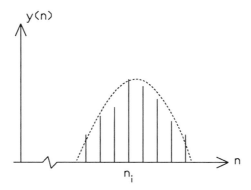

Figure 7.17. Least-squares fit to data.

$$y_i \simeq a + bx_i + cx_i^2 \qquad i = 1,2,3, \ldots, N$$

where the coefficients a, b, and c are to be determined. The squared error in the least-squares fit to the ith data point is $\epsilon_i^2 = (y_i - a - bx_i - cx_i^2)^2$. The coefficients a, b, and c that we seek are selected in such a way that they minimize the sum of these squared errors over all the N data points. The quantity to be minimized is therefore

$$e^2 = \sum_{i=1}^{N} \epsilon_i^2 = \sum_{i=1}^{N} (y_i - a - bx_i - cx_i^2)^2$$

Using standard techniques from calculus, we solve the following three equations

$$\frac{\partial e^2}{\partial a} = 2\sum_{i=1}^{N} (y_i - a - bx_i - cx_i^2)^1 (-1) \equiv 0$$

$$\frac{\partial e^2}{\partial b} = 2\sum_{i=1}^{N} (y_i - a - bx_i - cx_i^2)^1 (-x_i) \equiv 0$$

$$\frac{\partial e^2}{\partial c} = 2\sum_{i=1}^{N} (y_i - a - bx_i - cx_i^2)^1 (-x_i^2) \equiv 0$$

When terms are collected, the resulting equations can be written in the following convenient matrix form:

$$
\begin{bmatrix} \sum_i y_i \\ \sum_i x_i y_i \\ \sum_i x_i^2 y_i \end{bmatrix} = \begin{bmatrix} \sum_i 1 & \sum_i x_i & \sum_i x_i^2 \\ \sum_i x_i & \sum_i x_i^2 & \sum_i x_i^3 \\ \sum_i x_i^2 & \sum_i x_i^3 & \sum_i x_i^4 \end{bmatrix} \begin{bmatrix} a \\ b \\ c \end{bmatrix} \tag{7.14}
$$

A variety of techniques exist for solving for a, b, and c. The matrix inversion routine on a typical PC mathematics software package is entirely adequate. When these coefficients have been determined, the value x^* at which the least-squares fit achieves an extremum is given by $x^* = -b/(2c)$.

7.11.2. Additional Comments on Least-Squares Methods

This example of a second-order least-squares fit to data can be generalized to higher orders using the same minimization technique. Although the fit was used here to determine the location of the maximum, there are many other uses for the method. It can be used to average out noise errors in data or to determine the trend of a data set. In the latter application, a straight-line fit is often used. An additional attractive feature of least squares is that it can be applied when no information is available on the probability density function governing the errors in data. More sophisticated signal processing algorithms usually require this information.

The least-squares solution in Eq. (7.14) was developed on a component-by-component basis. This component-based approach can become quite tedious when a high-order least-squares fit is sought to a large data set. A more elegant and concise formulation of and solution to least-squares problems is to be found using matrix techniques where the (linear) relationship between the data vector Z and the unknown parameter vector X is $Z = HX + V$. The vector V represents noise present in the data. The error vector $e = (Z - HX)$ is formulated and the optimum solution $\hat{X}$ is obtained by minimizing the norm $e^T e = (Z - HX)^T(Z - HX)$. The result is $\hat{X} = (H^T H)^{-1} H^T Z$. The reader should verify that if this general solution is applied to the second-order fit problem, it leads directly to the solution of Eq. (7.14).

The weighted least-squares method is used when information is available on the relative accuracy of each measurement. This information is mathematically expressed in the form of a measurement covariance matrix R. A low-accuracy measurement is indicated by a large numerical value (variance) on the corresponding diagonal entry of R. Then, by selecting $(Z - HX)^T R^{-1}(Z - HX)$ as the norm to be minimized, the solution $\hat{X} = (H^T R^{-1} H)^{-1} H^T R^{-1} Z$ will have a built-in tendency to place less weight on the lower-accuracy data points and greater emphasis on the higher-accuracy points.

Matrix representations similar to those described here are typically used in software packages both to input data to the least-squares routines and to output the solution vector.

We have seen that the least-squares method can be applied in the absence of any information about the statistical properties of errors (noise) in the data. However, in certain occasions we need some information about the noise. For example, to design a filter to remove noise from data, we must know in what frequency bands the noise power is concentrated. The following technique is useful in obtaining this basic piece of information from the data.

7.11.3. Autocorrelation Function of a Data Sequence

The autocorrelation function of a data sequence is one of the fundamental procedures used in stochastic signal processing. The autocorrelation of the sequence $x(n)$ is defined as

$$R_x(m) \equiv E\{x(n)x(n-m)\}, \qquad m = 0, \pm 1, \pm 2, \pm 3, \ldots \qquad (7.15)$$

where $E\{\ \}$ is the mathematical *expectation* operator of probability theory. However, in practice we have access only to a finite segment of $x(n)$ of length N and hence cannot compute $R_x(m)$ using Eq. (7.15). We numerically compute instead, an estimate of the autocorrelation as follows:

$$r_x(m) \equiv \frac{1}{N} \sum_{n=0}^{N-1} x(n)x(n-m), \qquad m = 0, \pm 1, \pm 2, \pm 3, \ldots, \pm(N-1) \quad (7.16)$$

where N is the number of data points. Equation (7.15) is used to derive the properties of $R_x(m)$. If certain conditions known as *ergodicity* are satisfied, Eqs. (7.15) and (7.16) yield the same result. Ergodicity implies that statistical averages [such as Eq. (7.15)] are equal to time averages [such as Eq. (7.16)] carried out on a single sample function $x(n)$. These conditions are assumed to be satisfied in practice unless strong evidence exists to the contrary.

The simplicity of Eqs. (7.15) and (7.16) belies the mathematical sophistication of the procedures involved. While one must be well versed in the theory of probability and of random processes in order to fully grasp all that is implied by these equations, fortunately one need only understand a few important properties to use the technique to advantage. These are discussed next.

7.11.4.   Properties of the Autocorrelation Function

The autocorrelation $R_x(m)$ is an even function of m, that is, $R_x(-m) = R_x(m)$. This can be easily seen by observing that since the same sequence $x(n)$ is involved, delaying x by m units with respect to itself in effect is the same as advancing x by m units with respect to itself. $R_x(0) = E\{x(n)^2\}$ average power of $x(n) \geq R_x(m)$, $m > 0$. The autocorrelation has its maximum value at $m = 0$.

The next property is useful in analyzing data containing noise.

7.11.5. Autocorrelation and Power Spectral Density

The DFT of the autocorrelation function gives the power spectral density (psd) of the process $x(n)$. The psd tells us in what frequency bands the noise power is concentrated. The first step in deriving this important property is to note that the autocorrelation of $x(n)$ in Eq. (7.16) can be obtained by linearly convolving $x(n)$ with its own reverse [i.e., $x(-n)$] because

$$\frac{1}{N} x(n) \star x(-n) = \frac{1}{N} \sum_{n=0}^{N-1} x(n)x[-(m-n)] = \frac{1}{N} \sum_{n=0}^{N-1} x(n)x(n-m) = r_x(m) \quad (7.17)$$

We next observe that since the DFT can be used to linearly convolve two sequences (after appropriate zero padding), we can also use it to find $r_x(m)$. If $S_x(k)$ is the DFT of $r_x(m)$, then from Eq. (7.17) $S_x(k) = 1/N \, \mathrm{DFT}\{x(n)\}\mathrm{DFT}\{x(-n)\}$. Now $\mathrm{DFT}\{x(-n)\} = \sum_{n=0}^{N-1} x(-n)\exp(-j2\pi nk/N)$. Making a change of summation variable to $i = -n$, and recognizing that $x(i) = x(i + N)$, we get $\mathrm{DFT}\{x(-n)\} = \sum_{i=1}^{N} x(i)\exp(j2\pi ik/N)$. Finally, since $x(N) = x(0)$, we arrive at the important result that $\mathrm{DFT}\{x(-n)\} = \sum_{i=0}^{N-1} x(i)\exp(j2\pi ik/N) = X(k)^*$, where $X(k) = \mathrm{DFT}\{x(n)\}$. Therefore

$$\mathrm{DFT}\{r_x(m)\} \equiv S_x(k) = \frac{1}{N} X(k)X(k)^* = \frac{1}{N}|X(k)|^2 \text{ and}$$

$$r_x(m) = \frac{1}{N} \sum_{k=0}^{2N-2}|X(k)|^2 \exp\frac{j2\pi mk}{2N-1} \quad (7.18)$$

Note that in Eq. (7.18), $x(n)$ has been padded with $(N-1)$ zeros to eliminate circular convolution effects.

Setting m $= 0$ in Eq. (7.18), $r_x(0) = 1/N \sum_{k=0}^{2N-2}|X(k)|^2 =$ average power of $x(n)$. The quantity $|X(k)|^2$ is called the *periodogram* of $x(n)$ and is an estimate of the power spectral density. To summarize, the autocorrelation of an N-point data set $x(n)$ is obtained by first padding x with $(N-1)$ zeros and then taking the $(2N-1)$-point DFT to get $X(k)$. Next, find $|X(k)|^2$ and take the IDFT. The result is the $(2N-1)$-point $r_x(m)$.

Since the psd tells us in which frequency bands the noise power is concentrated, this can be useful in designing digital filters to separate a signal from noise. If the psd indicates that the noise power is concentrated in frequency bands that do not overlap with those of the signal, a digital filter can then be designed to pass the signal frequencies and eliminate the noise frequencies. Unfortunately, if the noise and signal occupy the same frequency bands, then a digital filter cannot be designed to separate them as it cannot differentiate between the two.

7.11.6.   Autocorrelation of a Signal Plus Noise

The autocorrelation function plays an important role in detecting the presence of a signal in noise. Suppose data $x(n)$ consists of a deterministic signal $s(n)$ plus additive zero mean noise $w(n)$. The noise is assumed to be independent of $s(n)$. Then $x(n) = s(n) + w(n)$ and $r_x(m) = E\{x(n)x(n-m)\} = E\{s(n)s(n-m)\} + E\{w(n)w(n-m)\}$. The "cross-terms" go to zero as $E\{s(n)w(n-m)\} = s(n)E\{w(n-m)\} = 0$ and so forth. Consequently, $r_x(m) = r_s(m) + r_w(m)$. Suppose now that $s(n)$ is a periodic signal such as a sinusoid whose period T is unknown. Then $r_s(m)$ will also be periodic with period T. Consequently, by inspecting $r_x(m)$ and determining the period of its periodic component $r_s(m)$, we can detect the presence of $s(n)$ in $x(n)$ and also estimate its period T. This is one of the many uses of the autocorrelation function, namely, to detect the presence of and to estimate the frequency of a periodic signal embedded in noisy data. This method is quite effective even when the data is so noisy that no signal can be observed from direct inspection of the raw data.

The autocorrelation function $r_x(m)$ of the signal plus noise also contains statistical information on the autocorrelation of the noise itself. As we see next, the degree of sharpness or "deltaness" of the noise autocorrelation $r_w(m)$ contains information on how "white" the noise is.

7.11.6.1.   White Noise

A noise sequence w(n) is defined as "white noise" if its autocorrelation $r_w(m)$ is the Kronecker delta function, $r_w(m) = K\delta(m)$, where K is some constant. Since the DFT of $r_w(m)$ is the constant K, white noise has a flat psd for all frequencies and consequently the average power $r_w(0)$ is infinite. While white noise clearly cannot exist in nature, it nevertheless is a very useful model for noise whose bandwidth is at least as large as that of the physical system (sensor, telecommunications channel, control system, etc.) that it impacts. The autocorrelation of noise samples can be used to establish how white the noise is. A sharply peaked value at $m=0$ indicates that the noise is white, whereas if the autocorrelation decays slowly toward zero, the noise is correlated; the more correlated the noise is, the longer the decay takes and the more low-frequency the noise is. If the noise has a nonzero mean value of μ, the autocorrelation decays to μ^2.

7.11.6.2. *Signal-to-Noise Ratio Enhancement Property*

An example of the signal-to-noise ratio enhancement properties of the autocorrelation function is given in Figure 7.18. The raw data shown in Figure 7.18a consists of 20 cycles of a 50-Hz sine wave to which is added zero-mean white Gaussian noise. The sinewave has an amplitude of 1, and the noise variance is 0.36.

The autocorrelation of the data is shown in Figure 7.18b. As anticipated, since the data contains a periodic signal, the autocorrelation also contains a periodic component. The spike occurring at the zero shift corresponds to the autocorrelation of the white noise. Recall that white noise, by definition, has an autocorrelation function that is an impulse. Thus we observe that the autocorrelation gives information about both the noise and signal components of the raw data.

Observe how relatively smooth the autocorrelation function is compared to the original data. The periodic component has the same period as that of the embedded sinusoid. Clearly, the period can be more accurately determined from the autocorrelation than from the raw data.

The period of the sinusoid is determined from the first peak (after the zero shift) in the autocorrelation function. Figure 7.18c is an expanded view of Figure 7.18b in the vicinity of zero shift, and it clearly shows the first peak occurring at 0.02 seconds. Thus we correctly conclude that the frequency of the embedded sinusoid is $f = 1/0.02 = 50$ Hz. If necessary, a linear least-squares parabolic fit about the peak could be employed to remove the slight perturbations due to the noise. Observe that the spike at zero shift has opened out slightly. This is to be expected as white noise cannot exist in practice, and consequently some broadening of the autocorrelation spike is to be anticipated. Nevertheless, the actual random noise is a very good approximation to white noise.

7.11.7. Additional Comments on the Autocorrelation Function

This discussion about autocorrelation techniques is by no means comprehensive. The intent is to familiarize the reader with the most basic concepts necessary to apply the technique; consequently, several subtleties have been ignored up to now. For example, since only a finite segment of data is involved, Equation (7.16) gives only an estimate of the autocorrelation defined in Eq. (7.15). Further, this estimate is known as the "biased" form of the autocorrelation. The bias asymptotically approaches zero as the data length $N \rightarrow \infty$, for those lag values $|m| < N$. It is shown in the references that the "unbiased" autocorrelation is obtained by using $1/(N-|m|)$ rather than $1/N$ as the normalizing constant in Eq. (7.16), where m is the lag value of the autocorrelation. However, for values of m approaching N, this normalization makes the variance of the "unbiased" estimate grow rapidly and consequently the "biased" form is often preferable. Clearly the more data that is available, the more accurate the resulting estimate is likely to be.

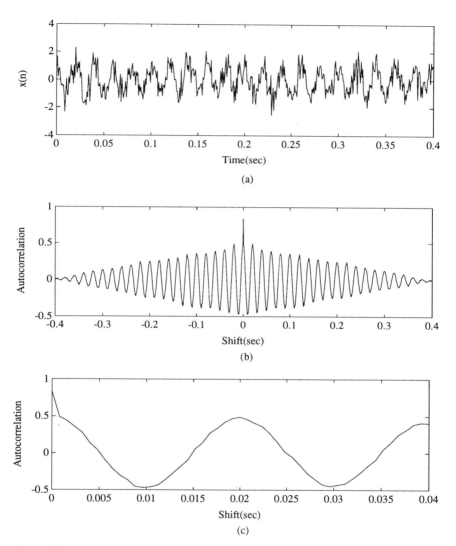

Figure 7.18. Autocorrelation of sinusoid plus white noise: (*a*) raw data; (*b*) autocorrelation; (*c*) expanded view.

311

The periodogram is not a consistent (in the statistical sense) estimate of the true power spectral density of $x(n)$. Nevertheless, the ease with which it can be computed using the FFT method has led to the development of methods to improve the estimate. These methods involve averaging periodograms obtained from consecutive sections of a long data record. One of the more commonly used methods, known as the *Welch method*, partitions a long data record into adjacent segments of length M, where M is a power of 2. Each segment is then tapered at the end by multiplying it by a window function such as the Hanning. The periodogram of each windowed segment of length M is computed and stored. Finally, the periodograms are added together and averaged to yield an estimate of the power spectral density that is statistically unbiased and consistent.

The technical details of these methods are discussed in the references, where averaging of consecutive estimates of the autocorrelation to get an improved estimate is also discussed. In any particular application one initially tries either Eq. (7.16) or (7.18) to estimate the autocorrelation. If these prove unsatisfactory, more data is collected and/or averaging of periodograms is used. If this is still unsatisfactory, more advance techniques from the general field of spectrum estimation may be required. This latter area is very extensive and mathematically sophisticated and lies beyond the scope of this material. Fortunately, the methods discussed in this section are adequate for many problems encountered in practice.

7.11.8.   Software for Computing the Autocorrelation and Periodogram

Most signal processing software packages include routines for implementing both the "biased" and "unbiased" estimates of the autocorrelation in Eq. (7.16). These packages also provide a routine for calculating the periodogram, including the averaging of consecutive estimates to improve accuracy as described in the previous section. Because the autocorrelation $r(m)$ involves both negative and positive lags m, this presents problems for software that cannot accommodate array addresses requiring negative or zero values. Several methods for dealing with this problem have been devised. In the first of these, since $r(-m) = r(m)$, only the nonnegative lags are computed with the zeroth lag appearing in the first array location. Another method is to generate all $(2N-1)$ lags of an N-point data segment. The negative lags appear in the first $(N-1)$ array locations and the zeroth lag is in the Nth. The positive lags fill out the remaining $(N-1)$ locations. Minor variations of these two methods also exist. Consequently, when using a given software package to compute the autocorrelation, it is imperative that the format for presenting the output values be clearly understood.

Having discussed signals and noise in a general setting, we now focus attention on noise generated by sensors.

7.12.   MODELING AND REDUCTION OF SENSOR ERRORS

There are two basic types of sensor errors. The first consists of the zero-mean random variety and the second is the consistent, bias-type error. We consider each type separately, starting with the first.

7.12.1.   Zero-Mean Random Sensor Errors

This type of error, or "noise," is sometimes caused by the finite resolution capability of the sensor. In the case of an analog sensor, it may arise from errors in precisely reading the position of the indicator such as a needle on a graduated scale, for example. One reading may be slightly high, and another reading may be slightly low. Another source may be a nearby disturbance causing the sensor to give an incorrect reading. Countless other situations can be envisioned as causes of sensor errors.

The common characteristic of these various errors is that they are all zero-mean. This implies that the impact of the errors can be minimized by some appropriate averaging technique that uses positive and negative errors to offset each other. The least-squares method discussed earlier is one of the most common techniques used for averaging out the effects of noise in data. Its appeal lies in its simplicity and ease of application and it has a major advantage in that it can be applied without regard to the probabilistic distribution governing the noise.

The Gaussian distribution (also known as the *normal* distribution) is one of the most widely used probabilistic models for sensor noise. The justification for this lies in the theoretical underpinnings of the central-limit theorem of probability theory. In simplified terms, this theorem states that when several random disturbances act on a system (sensor) in an independent manner, the combined effect of these disturbances tends to have a Gaussian distribution even if the distributions governing the individual disturbances are not Gaussian. This is a very powerful theorem, which has been found to be valid in a vast number of practical applications spanning the range from single sensors to complex aerospace systems.

Many sophisticated techniques based on the Gaussian distribution have been developed for the reduction of noise in sensor data. These techniques, which use mathematical models of dynamic systems, optimization methods, and the theory of probability and random processes, are beyond the scope of this material. However, an appreciation of the techniques involved can be obtained from the following simple example involving the optimal combination of measurements from two sensors having different accuracies.

7.12.1.1. Combining Measurements from Two Sensors
of Different Accuracies

Suppose that sensors A and B, measuring a given quantity, produce measurements Z_a and Z_b, whose errors have zero mean and variances σ_a^2 and σ_b^2, respectively. Note that these variances are usually provided by the manufacturer of the sensor; indeed, the smaller the variance, the more accurate and consequently the more expensive the sensor is. Suppose further that $\sigma_a^2 \ll \sigma_b^2$, that is, the measurements from A are much more accurate than those from B. We now pose the following question. Since B is of much lower accuracy than A, should it be discarded and only sensor A's measurements be used, or can the measurements from A and B be combined to produce a fused measurement that, *on average*, is statistically superior to either of the two original ones? Statistical signal processing methods provide a "yes" answer to the latter option.

Briefly, the procedure is to form a weighted (or fused) measurement $Z = \alpha Z_a + \beta Z_b$, where the constants α and β are to be determined. Since Z is also to have a zero-mean error, we require that $\beta = (1-\alpha)$. Consequently, $Z = \alpha Z_a + (1 - \alpha)Z_b$. Assuming that the errors in Z_a and Z_b are statistically independent, the variance of Z is $\sigma_z^2 = \alpha^2\sigma_a^2 + (1-\alpha)^2\,\sigma_b^2$. The final step is to use mathematical optimization methods to select that value of α that minimizes σ_z^2. This will guarantee that the fused measurement Z, *on average*, is as accurate as possible (has a minimum variance) within the limits imposed by the accuracies of the two individual sensors. Using conventional techniques from calculus, we find that the optimum values are $\alpha = \sigma_b^2/(\sigma_a^2 + \sigma_b^2)$ and $\beta = \sigma_a^2/(\sigma_a^2 + \sigma_b^2)$. The variance of Z is $\sigma_z^2 = 1/(\sigma_a^{-2} + \sigma_b^{-2})$. It is left as an exercise for the reader to verify that $\sigma_z^2 < \sigma_a^2$ and $\sigma_z^2 < \sigma_b^2$. Consequently, the fused measurement has, on average, a higher accuracy (smaller variance) than either of the two original measurements.

The reader is urged to examine the limiting form of both the fused measurement Z and its variance σ_z^2 as σ_b^2 approaches infinity. Are these limiting forms what one might expect?

7.12.2. Bias-Type Sensor Errors

Unlike the zero-mean random errors just described, this type of error is consistent from one measurement to the next and generally cannot be eliminated by averaging methods. It is often caused by a miscalibration of the sensor, which causes readings to be biased high or low. The conventional method for dealing with this error is to periodically take the sensor out of use and recalibrate it or to calibrate it immediately prior to use and trust that the calibration will remain valid for the duration of use. Emerging "smart sensor" techniques are building into the sensor an ability to perform self-diagnosis. The sensor recalibrates itself, often using a

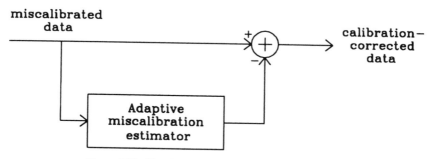

Figure 7.19. Signal processing-based recalibration of data.

high-tolerance specimen provided for the purpose, or indicates an impending malfunction to a monitoring facility.

Another method of sensor self-calibration involves applying advanced signal processing techniques to the data stream produced by the sensor. As shown schematically in Figure 7.19, these techniques adaptively estimate the miscalibration, or bias, injected by the sensor into the data stream and then correct the data prior to passing it downstream for subsequent processing. The advantages of this signal processing-based approach are that it is automatic, is continuously available, and eliminates the time and expense of taking the sensor off line for recalibration. It can also compensate for an additive bias that changes with time. This change may be gradual, as might result, for example, from the sensitivity of some sensor component to fluctuations in the ambient temperature. Or it might be sudden, as might arise from an internal sensor component undergoing an intermittent change in value.

In order to estimate an additive miscalibration using signal-processing-based techniques, a good mathematical model must first be developed for the underlying signal. For those signals having an analytical description, such as sinusoids or exponentials, such a mathematical model is easily developed. The changing miscalibration can be successfully modeled by an additive first-order Markov process. A higher-order model incorporating a velocity term may be necessary to enhance convergence in the case of a rapidly varying miscalibration. The signal and miscalibration models are combined to form the state variable model of the overall system. An adaptive estimation algorithm is then used both to estimate the bias from the data stream and also to reduce any zero-mean noise that might be present. There are several forms of the adaptive estimation algorithm, ranging from the simple alphabeta filter to the more sophisticated Kalman filter algorithm. These are discussed in detail in the references, particularly in the last four entries.

If the method is to be successful, however, the crucial test of *mathematical observability* of the state variable model must be satisfied. If this test is failed, or indeed if an accurate mathematical model for the underlying signal is not available, then direct recalibration may be the only alternative. Unfortunately, the issue of mathematical observability, whose function it is to determine whether the miscalibration can be successfully extracted from the sensor data stream, cannot be adequately discussed here. Suffice it to say that the most potentially troublesome situation involves a sensor with an additive miscalibration that does not change with time, measuring a signal that also does not change with time. The observability test is very likely to be failed under these conditions. Paradoxically, when either the signal being measured, or the miscalibration itself, or both, are changing with time, conditions are more favorable for signal-processing-based automatic recalibration of the data stream than when both quantities are constant with time.

If the miscalibration is multiplicative instead of additive, as might arise from a sensor with an incorrectly calibrated gain, for example, the problem is much more complex. To tackle this type of problem using signal processing, one must inevitably resort to nonlinear estimation techniques.

7.12.3. Additional Comments

Because of the advanced concepts involved, this discussion of sensor errors is, of necessity, very brief. Nevertheless, it is an extremely important area of signal processing because all sensors inevitably introduce some noise into the data stream. Consequently, the last four entries in the "Suggested Further Reading" section are provided for those wishing to delve further into this important and fascinating section of signal processing. Collectively, they accommodate both those seeking an introductory exposition of the material and those who need a comprehensive theoretical discussion of the more advanced topics.

7.13. CONCLUSIONS

The field of signals, signal processing, and indeed system modeling is very vibrant and in a state of continual evolution. A brief excursion such as this one into an area as extensive as signal processing must of necessity strike a delicate balance between the depth and the breadth of the material discussed. This is particularly true when it is assumed that the reader has had little or no prior exposure to the material in question. With this in mind, the list of references, although far from exhaustive, was selected to enable the reader to pursue further the topics touched on here.

It is the author's hope that the pedagogy of the presentation provides for the uninitiated a foothold in this somewhat esoteric material. This foothold, it is hoped, provides a solid support for the beginner to both apply signal processing methods to data and also pursue a deeper understanding of the theoretical issues involved. In addition, the information presented here should enable the reader to intelligently select from and use the signal processing software packages that now appear on the marketplace in ever-increasing numbers.

SUGGESTED FURTHER READING

Digital Signal Processing

Brigham, E. O., *The Fast Fourier Transform,* Prentice-Hall, Englewood Cliffs, NJ, 1974.

DeFatta, D. J., Lucas, J. G., and Hodgkiss, W. S., *Digital Signal Processing: A System Design Approach,* Wiley, New York, 1988.

Oppenheim, A. V., and Schafer, R. W., *Digital Signal Processing,* Prentice-Hall, Englewood Cliffs, NJ, 1975.

Proakis, J. G., and Manolakis, D. G., *Introduction to Digital Signal Processing,* Macmillan, New York, 1988.

Terrell, T. J., *Introduction to Digital Filters,* 2nd ed., Macmillan Education, London, 1988.

Statistical Signal Processing

Brown, R. G., *Introduction to Random Signal Analysis and Kalman Filtering,* Wiley, New York, 1983.

Gelb, A., *Applied Optimal Estimation,* MIT Press, Cambridge, MA, 1974.

Larson, H. J., and Shubert, B. O., *Probabilistic Models in Engineering Sciences* (Vols. 1 and 2), Wiley, New York, 1979.

Shanmugan, K. S., and Breipohl, A. M., *Random Signals: Detection, Estimation and Data Analysis,* Wiley, New York, 1988.

INDEX